2nd Annual AIC Conference Anthology

2nd Annual AIC Conference Anthology

Edited by:

Austin Mardon
Catherine Mardon
Zachary Schauer
Jessica Jutras
Zachary Irving
Alyssa Kulchisky

Typeset and Cover Design by Kim Huynh

ISBN 978-1-77369-633-1
EBook 978-1-77369-636-2

Golden Meteorite Press
103 11919 82 St NW
Edmonton, AB T5B 2W3
www.goldenmeteoritepress.com

TABLE OF CONTENTS

TABLE OF CONTENTS

INTRODUCTION

The Antarctic Institute of Canada, formed in 1985 by former Antarctic Researcher, Austin Mardon, is a non-profit organization based in Edmonton, Alberta, Canada. However, the Institute's name is somewhat a misnomer. Though the Antarctic Institute of Canada started with and evolved from Dr. Mardon's research in Antarctic geography, it has quickly grown to encompass almost any area of contemporary research. It's accordant publishing company, Golden Meteorite Press (also founded by Dr. Mardon), publishes books in a plethora of genres, including natural sciences, history, psychology, theology, self-help, biography, translations, narrative, and literary criticism. Additionally, Dr. Mardon has published articles in a vast array of academic journals that reflects this diversity, since his CV is littered with astronomical studies, medical sciences, and technological developments, to name a small sample.

It is difficult to explain why the Antarctic Institute of Canada, seeming to be focused on Antarctica, has developed into such a large project, but, because the present anthology reflects this plurality, an explanation, however brief, is in order.

In 1992, Dr. Mardon, having returned from a journey in Antarctica and a trip to Moscow wherein he was incarcerated under the pretense of being an undercover spy, was diagnosed with schizophrenia. Before this diagnosis, Dr. Mardon was an avid researcher who never strayed more than a few metres from his pen. However, as schizophrenia tends to do, he was stripped of his ability to read and write with the same vigour and consistency that he once could. At that time, Dr. Mardon was primarily a geographer who, though interested in other fields, mostly stayed in his own academic lane, so to speak. The harsh nature of schizophrenia, combined with an overwhelming social stigma, lack of efficient medication, and permanent disabilities caused by his Antarctic adventure, left Austin abandoned by his colleagues, his family, and, it seemed, his muse.

Without the power to pursue the academic life he had strove for prior to his diagnosis, Dr. Mardon longed for a way to stay connected to his scholarly past while foregoing the apathy that often comes attendant with schizophrenia. He put his time into volunteering, recuperating his life, and searching for outlets for his mind, which was always teeming with ideas for books, articles, and research. So, Dr. Mardon decided to try his hand at opening his own publishing firm with the hopes that if he could not perform the research that he had loved so much, at least he might help others continue that legacy and begin to make a name for themselves. The project, maybe crazy to some outsiders (Dr. Mardon considers his craziness to, sometimes, be a compliment), turned out to be a great success.

Dr. Mardon began to receive the Canada Summer Jobs grant, which allows employers in Canada to hire students and youth for summer-long internships. With the help of these grants, Dr. Mardon was able to hire students and put them to the task of writing and, with the help of his newly established publishing company, he had a forum to publish their works. The Canada Summer Jobs program was an experiment that started on a small scale, yielding a few grants to a few students that resulted in a few books. Though small to some outsiders, especially from large research institutions and universities, Austin's recovered ability to engage in academics was, for him, a return from the ashes and darkness of severe mental illness. He never recovered from schizophrenia, but found an opportunity to thrive despite its aggression.

Fast forward to 2021, and the Antarctic Institute of Canada has now grown a life of its own far beyond the expectations that Dr. Mardon held. Instead of a few thousand dollars, the Institute received $2 million in funding through different programs so that it could foster hundreds of student researchers, designers, audio engineers, and program managers that were able to produce hundreds of books, articles, and audiobooks in the course of a few summer months. Given the scale of the programs in recent years, the subject matter has expanded in tandem. The present book will give a reader an idea of the grandeur that the Institute has taken on in recent years, and will showcase the variety of methods, disciplines, authors, and genres that the Institute circumscribes.

Dr. Mardon has always had a clear purpose in mind throughout all the seemingly disconnected research that appears under his curriculum vitae: to encourage and empower new students, including those who are disadvantaged, to pursue their research. Dr. Mardon has given a platform to countless students from all walks of life. The AIC Conference, founded by the Antarctic Institute of Canada intern, Zachary Schauer, accords exactly with that principle. The conference was designed to allow students, some still in high-school, some Master's degree alumni, to present their ideas around like-minded students who thirst for knowledge. As such, there is a breadth of different subjects presented at each conference from researchers across Canada and around the world.

It is our hope that this anthology is more than just another opportunity for students to publish material and stack citations on their resumes, CVs, and Graduate applications, but that it will also inspire readers to see that the power to pursue their own research interests lies within them. There are always people willing to hear novel ideas and there are always forums where one can share those ideas. The 2021 AIC Conference was just one of those places, which, we hope, will continue to flourish as time progresses.

As a note for readers, not all the presenters at our conference wished to have their works published in this volume, and so, in lieu of those presenter's articles, we have acknowledged their contributions in this book by providing a short biography so that readers can look up those authors at a later time. Further, some articles could not be published due to copyright restrictions, so, in those cases, the authors were given the chance to submit stand-in articles that substitute for their conference articles, and we have printed those articles here.

It is a limitation to all introductions that the full extent of the author's purpose, along with their history and accomplishments, cannot be covered. Curious readers may always contact the Antarctic Institute of Canada directly or read up about Dr. Austin Mardon through the various biographies and online articles that debrief one more thoroughly. We hope that readers enjoy the wonderful research submitted by our presenters and that the presenters are gratified by a print book that bears their work.

Per ardua ad astra.

Zachary Schauer.

ACKNOWLEDGMENTS

The Antarctic Institute would like to thank the presenters of the conference for taking the time to attend our conference and engage with other students in such a way that awakens the academic spirit. We also thank other conference organizers, without whom the conference would be impossible, including Jessica Jutras, Zachary Irving, and Alyssa Kulchisky. Special thanks must always be given to Catherine Mardon, Dr. Austin Mardon's wife, who not only gave a keynote speech, but continuously supports us throughout all the hurdles, triumphs, frustrations, and successes that the Institute faces. We also thank our volunteer organization, Sharpen the Quill for producing volunteers who write like they are actually getting paid. Finally, we would like to thank Riipen's LevelUp program and Canada Summer Jobs, which funded many of the students as they were producing their conference materials, and #RisingYouth for funding the conference itself. It is a joy to have so many avenues of support allowing events like these to take place.

DAY 1

PANEL 1: REFLECTIONS ON AGING

The Search for Eternal Youth

Madeline Langier

Increased longevity is to challenge nature by increasing the human lifespan past what may be considered natural. Hence, increased longevity or the ability to acquire eternal life has long been associated with the divine or the mystical. Today, modern science has begun to provide a taste of what only the Gods have had, eternal life. Advancements in technology, medicine and intelligence have bestowed the human race with the gift of a bit more time on earth. Currently, cryogenic freezing and stem cell research are in the forefront of popular cosmetic anti-ageing and longevity treatments affecting the human lifespan. However, caloric reduction and research to slow or stop cell decay are widely considered the most effective and innovative methods to accomplish increased longevity. Before modern technology, cryogenic freezing for cosmetic anti-ageing was used to various ends (although not for freezing the body for a century to unfreeze when science has found a cure for ageing). Rather, freezing as an anti-ageing treatment targets the outer layer of the skin to impact deep layers of collagen and is known as cryotherapy. Medical professionals have been using cryotherapy for nearly 100 years to treat common skin conditions (Andrew, 2004). Specifically, cryotherapy is used to treat non life threatening skin conditions and is referred to as cryosurgery by modern medicine. In cryosurgery, liquid nitrogen is applied to the skin using a cotton tip or a liquid nitrogen spray (Andrew, 2004). Frigid temperatures are used to freeze abnormal skin tissue, and research has found it may be effective in treating acute injuries (Hubbard et al., 2004).

In addition, cryosurgery helps to treat skin infections such as warts and is incredibly useful in treating sun-damaged skin and related skin problems (Andrew, 2014). Cryogenic therapy has reached the mainstream consumer market and is used as a tool to dispel ageing and fat cells to maintain the appearance of adolescents and virility. However, cryotherapy research is not necessarily marketed in the medical field as an anti-ageing treatment. While it is used cosmetically to treat superficial elements of skincare cryogenic therapies, it assists in longevity because it increases the quality of life. Therefore cryotherapy is a unique tool with the potential to contribute to the well-being and superficial elements of longevity.

Before modern science, various metaphysical elements were speculated on to address ageing. The earliest known record mentioning the search for eternal life is found in the ancient Mesopotamian poem about the King and demigod, Gilgamesh. An English archaeologist and explorer, Austen Henry Layard, discovered the Epic of Gilgamesh in 1849 CE. The epic, a long poem describing the various perilous adventures undertaken by the heroes of days long past, dates back to the Mesopotamian era, which roughly spanned from 500 BCE and 400 CE. Layard discovered the epic on twelve stone tablets in the ruins of the ancient library Ashurbanipal in the Sumerian language, although it was later translated into Akkadian, the native tongue of the Mesopotamian empire (Mark, 2021). Most notable is that the epic features Gilgamesh and his human struggles against death, loss and the seemingly meaningless human existence. While the epic is widely known as a legend containing many mystical elements, many scholars believe Gilgamesh to be a real ancient king who ruled the city of Uruk 100 years before the creation of the stone tablets.

In the epic, Enkidu, a wild man tamed by a temple concubine, is brought to the city of Uruk. Enkidu challenges Gilgameshto a battle and is overcome, after which the two warriors pledge themselves to each other in a pact of eternal friendship. Further, Gilgamesh's mother adopts Enkidu as her own. The pair share an intimate bond. Eventually Enkidu dies, and Gilgamesh becomes lost, leading him to search for the cure to death. In his despair, Gilgamesh ventures to the Land of Night and the Waters of Death, where he finds the ancient man Utanapishtim, the only human being to survive the Great Flood who was afterwards granted immortality (Mark, 2021). Utanapishtim tells Gilgamesh the story of how he was warned by the god Ea of the coming deluge, followed his command to build an ark, place assorted animals inside, and save himself and his family from death and humanity from extinction. He then tells Gilgamesh that eternal life will be granted if the great King can stay awake for the next six days; he is unable to. In a second attempt for the gift of eternal youth, Gilgamesh found and aimed to retrieve a magic plant that would make one young again. According to the legend, a snake ate the plant while Gilgamesh slept. Simply put, the magic plant gave the snake its chance to rejuvenate.

Having failed to win immortality, Gilgamesh is brought back to Uruk by the ferryman Urshanabi where, once home, he writes down his great adventure (Mark, 2021) According to the historian D. Brendan Nagle, the poem deals with human problems such as sickness, death, fame, and unattainable desires. Furthermore, the epic could be considered a metaphor for Mesopotamia's own heroic struggle to resist decay and leave a name for itself among the peoples of earth (Mark, 2021). Regardless of the assumptions that might be made about the meaning of the story, ultimately Gilgamesh fails his epic mission to attain eternal life.

While the Epic of Gilgamesh encapsulates the search for longevity to evade death, others have sought to avoid ageing altogether. The search for eternal life is not the same as the desire for eternal youth in that eternal life includes the functioning of the body for eternity whereas ageing is a superficial element typically connected to what we can see. The Byzantine Empress Zoe Porphyrogenita effectively represents many peoples' continuous struggle for eternal youth rather than the desire for eternal life. Among the living from 978 to 1050 and a beautiful empress, Empress Zoe Porphyrogenita was blonde, with bright white smooth skin well into her sixties. Moreover, she was described as child-like, sharing features with a very young girl (Panas et al., 2012). Historians describe the Empress as entirely committed to formulating cosmetic extracts. She installed a laboratory in her private quarters in the palace where she spent most of her time manufacturing drugs and perfumes (Panas et al., 2012). Various sources suggest she at the time she spent was almost to create potions which she may have been used to poison each of her husbands.With this in mind, she married three times, at the ages of fifty, fifty-six, and sixty-four; the third time the bridegroom was roughly twenty years old. According to the eyewitness of those times, Byzantine historian and courtier Michael Psellos (1018 - c. 1082 CE), Empress Zoe had blonde hair into old age. Further, Psellos writes: "although she had already passed her seventieth year, there was not a wrinkle on her face. She was just as fresh as she had been in the prime of her beauty" (Panas et al., 2012).

In order to appear youthful, it is likely Empress Zoe dyed her hair and spent a great deal of time applying cosmetics to lighten her skin. Two prominent physicians, Alexander of Tralles (6th century) and Paul of Aegina (7th century), refer in their works to the methods of dying hair. In order to dye the hair blond or red, herbs such as myrrh, lime, saffron, sandarach and golden herb are mixed with a fatty substance and applied to the head for one or two days (Panas et al., 2012). In order to achieve the

9

most prominent feature of classic Greek and Roman beauty, pure white skin, native Greek, Roman and Byzantine women who were not naturally fair-skinned applied whitening make-up. In order to achieve the desired skin tone, the make-up usually contained white lead and chalked powder. Although aware of the poisonous attributes, women were willing to overlook the harm to obtain beauty (Panas et al., 2012).

It is worth nothing that the concept of youth often aligns very closely with beauty. The Byzantine women were willing to undergo painful and poisonous treatments to maintain their beauty, which is not entirely unlike treatments of more contemporary societies. During the life of Empress Zoe, the treatment to obtain eternal youth included seemingly extreme treatments using lead and chalk. The anti ageing methods of today, such as the previously mentioned cryotherapy, which researchers use to freeze off abnormal cells was eventually transformed into a tool to preserve youth (Hubbard, 2014). Without a doubt, such treatments required a great deal of research and trial and error in order to accomplish the lofty longevity goals of the modern world. Indeed, there are entire television shows following the result of botched surgeries and attempts to chase and capture eternal youth, a practice not dissimilar to the Mediterranean women of the Byzantine era. Each process involves unnatural alterations and has the potential to go horribly wrong. Historically, the appearance of youth well into old age held many parallels to longevity, and research conducted on Byzantine medical practices found cosmetic medicine made up a huge portion of the material (Panas et al., 2012). However, the obsession with beauty and eternal youth can not be considered the same as the desire for increased longevity. For Empress Zoe, the quest for eternal youth dominated her life, which is not the same as a desire for immortality or, more realistically, a life span much longer than average.

The search for longevity in the epic of Gilgamesh and the search for eternal youth by Empress Zoe are but two of many searches throughout history of those wanting to escape the effects of aging. Alexander the Great, who conquered many new lands during the fourth century BCE, is believed to have devoted time searching for a "river of paradise" to prevent ageing. In the fifth century BCE, the Greek historian Herodotus described a fountain responsible for extending the lives of the Macrobians, who lived in northern Africa in ancient times (Rohland, 2020). Beginning in the twelfth century CE, Europeans circulated a legend of a king named Prester John, supposedly ruling a kingdom containing not only a fountain of youth but also a river flowing with gold. Tales of hot springs capable of repairing injuries and reversing the ageing process have persisted in Japan for centuries (Rohland, 2020). This long tradition of attempting to escape the clutches of age and death persist today.

In addition to the anecdotes described above, one particularly well known legend is that of the Fountain of Youth. The Fountain of Youth is often initially associated with Juan Ponce de León, an explorer of parts of the Caribbean in the late fifteenth and early sixteenth centuries on behalf of Spain. Native inhabitants of various Caribbean islands reportedly shared accounts of a body of water (a river, a waterfall, a spring) that could transform adults into children. One alleged location was a lost land called Bimini, located north of Cuba and Haiti. According to legend, Ponce de León set out to search for this land to find the magical Fountain of Youth. In the process, he discovered present-day Florida, which he claimed for Spain and proceeded to explore. Some accounts claim that he first came ashore in St. Augustine, Florida, the oldest continuously inhabited European settlement in America. Other accounts indicate that he came ashore farther south in Melbourne, Florida. Regardless, Florida boasted many natural springs, and Ponce de León believed that one of them could be the Fountain of Youth. While searching for gold to claim on behalf of Spain, he also searched for the legendary fountain (Rohland, 2020).

Modern history scholars generally agree that Ponce de León's quest for the Fountain of Youth is more fiction than fact. Contracts between Ponce de León and the Spanish crown do not mention the Fountain of Youth, and few other records written by Ponce de León survive. Historians know that Ponce de León anchored his ship off the eastern shore of Florida on April 2, 1513, and began exploring land the next day. Eight years later, in 1521, he returned to Florida to attempt to build a colony. A Native American shot Ponce de León in the leg with an arrow, and the explorer died in Cuba shortly after that.

So, why the association between the explorer and the legend? One of the first people to connect Ponce de León with the Fountain of Youth was Spanish historian Gonzalo Fernández de Oviedo y Valdés. Oviedo and Ponce de León did not get along. Oviedo's 1535 account of Ponce de León's explorations set out to discredit the explorer by claiming that he had foolishly searched for the Fountain of Youth in an attempt to cure sexual impotence. Later, historians continued to connect Ponce de León to a quest to find the Fountain of Youth. In the nineteenth century, when Spain ceded Florida to the United States, the legend of Ponce de León's quest became entrenched in history and even found its way into history textbooks. In the twenty-first century, the legend persists, if only because people still dream of a way to ensure their perpetual youth and vigour (Rohland, 2020).

The legend of Ponce de León's quest for the Fountain of Youth has lasted for centuries. Indeed, several towns in Florida purport to be the home of the "real" Fountain of Youth. One of the most well-known is St. Augustine. The fountain's association with "America's oldest city" began in the 1870s when a real estate promoter dubbed a small stream "Ponce de León Spring" and claimed that it was the Fountain of Youth. Today, St. Augustine is home to the Fountain of Youth Archaeological Park (Rohland, 2020). Thousands of tourists visit the park each year to learn about the Spanish settlers who founded the city and the Timucua people who lived there for nearly three thousand years before the settlers' arrival. For many, a highlight of the park tour is tasting water from a stone well identified as "The Fountain of Youth," but most visitors report that the water tastes a bit like rotten eggs (Rohland, 2020).

Modern historians generally agree that the Fountain of Youth and Ponce de León's quest to find it are the stuff of fantasy. Not to mention it had little impact on the search for eternal youth (Rohland, 2020). Less than a hundred years after Ponce de León's explorations for the Fountain of Youth, gruesome news emerged of the Hungarian countess Elizabeth Báthory, another said to have pursued prolonged youth (Pallardy, 2020). Married for the first time at eleven and again at fifteen, Báthory gave birth to five children, but only three survived to adulthood. Further records show her second husband, a skilled soldier, died in 1604, leaving Báthory with an impressive estate (Biography.com, 2020, Pallardy, 2020). After the death of her husband, the rumours began to circulate, suggesting Báthory used her status to lure young virgins to her estate. Specific accusations were made that Báthory, jealous and filled with hate, would coat her victims with honey and leave them outdoors for the insects to feast (Pallardy, 2020). Further tales recall the countess sometimes tortured girls by driving needles into their fingers, cutting their noses or lips or whipping them with stinging nettles. She would bite shoulders and breasts, as well as burning the flesh, including the genitals, of some victims (Pallardy, 2020).

Apparently, in addition to various torture methods, legends speculate Báthory would bathe in the blood of virgin maidens, in hopes of discovering her lost youth. Although there is no hard evidence to suggest Báthory bathed in the blood of virgins, written historical records recall Báthory's arrest in 1611; the trial documents reaffirm these accusations made against her (Biography.com, 2020, Pallardy, 2020). The legend of the Blood Countess should be considered carefully as the first mention of Bathory's blood baths came a hundred years after her death. Further, modern scholarship has questioned the allegations' veracity. Primarily due to the fact that Báthory was a powerful woman, made more so by inheriting control of her husband's wealth. Further evidence suggests the Báthory family, aside from the countess, cancelled a large debt owed by a fellow noble, to the Báthorys in exchange for allowing them to control her in captivity. All of which suggests that the accusations were politically motivated slander that allowed relatives to appropriate her lands (Biography.com, 2020). While the countess was a wealthy and powerful noblewoman accused of horrifying age-defying rituals it is unlikely the legends reflect reality. (Pallardy, 2020) Then again, historical records show there may be some truth to the accusations. In 1602 a priest wrote a letter that discussed the excessive cruelty exhibited by Báthory and her husband towards their servants. The testimony against Báthory likely included true tales about how harshly she acted with lower classes. Such acts were not illegal at the time — Báthory was only punished because her victims were said to have included noblewomen — but would still make Báthory responsible for many ruined lives (Biography.com, 2020). Báthory would go on to be known as the Blood Countess and is often considered the inspiration for the Legend of Count Dracula. The legend of the virgin blood baths continued.

Today, the quest for longevity typically coincides with the much more attainable search for eternal youth. In late December of 2016, scientists at the Salk Institute for Biological Studies announced that they had found a way to reverse the signs of ageing in mice (Ocampo et al.). The process, called cellular reprogramming, extended the lives of the mice involved in the study and allowed them to live longer without experiencing the usual signs of ageing. For example, the reprogrammed mice lived about thirty percent longer than mice that had been allowed to age normally. The reprogrammed mice appeared younger and healthier and demonstrated improved organ function. The study has given scientists hope that they may one day achieve similar results in human subjects, although such experimentation would not likely occur for another decade. There have been successful studies on cellular reprogramming that has prompted some researchers to call cellular reprogramming a real "fountain of youth." (Rohland, 2020). Today, adding life to years cannot be viewed simply as a medical issue because the quality of a prolonged life depends on cognitive, behavioural, psychological, and social processes (Lang et al., 2019).

References

Andrews, M. D. (2004). Cryosurgery for Common Skin Conditions. American Family Physician University School of Medicine, 15;69(10):2365-2372.

Biography.com (2020). Elizabeth Bathory Biography. The Biography.com website. A&E Television Networks. https://www.biography.com/crime-figure/elizabeth-bathory.

Rohland, L. (2020). Fountain of Youth. Salem Press Encyclopedia.

Lang, F. R., & Rupprecht, F. S. (2019). Motivation for Longevity Across the Life Span: An Emerging Issue. Innovation in Aging, 3(2). https://doi.org/10.1093/geroni/igz014

Mark, J. J. (2021, May 15). The Eternal Life of Gilgamesh. World History Encyclopedia. https://www.worldhistory.org/article/192/the-eternal-life-of-gilgamesh/.

Panas, M., Poulakou-Rebelakou, E., Kalfakis, N., & Vassilopoulos, D. (2012). The Byzantine Empress Zoe Porphyrogenita and the quest for eternal youth. Journal of Cosmetic

Dermatology, 11(3), 245–248. https://doi.org/10.1111/j.1473-2165.2012.00629.x
Pallardy, R. (2020). Elizabeth Báthory. Encyclopedia Britannica. https://www.britannica.com/biography/Elizabeth-Bathory

Ocampo, A., Reddy, P., Martinez-Redondo, P., Platero-Luengo, A., Hatanaka, F., Hishida, T., Li, M., Lam, D., Kurita, M., Beyret, E., Araoka, T., Vazquez-Ferrer, E., Donoso, D., Roman, J. L., Xu, J., Rodriguez Esteban, C., Nuñez, G., Nuñez Delicado, E., Campistol, J. M., Izpisua Belmonte, J. C. (2016). In Vivo Amelioration of Age-Associated Hallmarks by Partial Reprogramming. Cell, 167(7), 1719–1733.e12. https://doi.org/10.1016/j.cell.2016.11.052

PANEL 2: RECENT IMPACTS OF COVID-19

Gambling Addictions and the COVID-19 Pandemic

Yash Joshi

Abstract

Gambling is a leisure activity for many individuals and has been for a very long time. It has a high probability to lead to an addiction because it impacts dopamine in the body and creates a sensation of pleasure. Addictions related to gambling have always been troublesome and a concern in many families even prior to the COVID-19 pandemic. The pandemic resulted in significant financial uncertainty for many, which led to a multitude of people exploring gambling. The closure of in-person casinos caused the shift to online betting, where frequent high-risk bettors took a reserved approach, yet there was an overall increase of total players. Mental health concerns related to gambling have emerged as many of those who participated in gambling during the lockdowns reported symptoms of anxiety and depression. Increased substance use due to the pandemic and gambling are also an issue of concern. Overall, the pandemic has left many exposed to the risks of gambling addictions, which is a situation that needs to be monitored. This review compiles and investigates sources from literary research done regarding gambling and the COVID-19 pandemic.

Introduction

Gambling is a popular activity in the adult world which often develops into an addictive behaviour for many. There is a big appeal to win a large sum of money, and just as much heartbreak when one loses that money. The chance of massive rewards gives it a highly addictive nature, which causes significant issues in the daily lives of many individuals. The COVID-19 pandemic shut down many casinos and forced people within the confines of their home, yet they were still connected to the world of gambling. Online casinos and sports betting have increased since the start of the pandemic, meaning that the industry itself is not facing any major losses. Recent data and patterns suggest that an increasing number of individuals are trending towards gambling, which is becoming a concern. There are certain socioeconomic reasons as to why this surge is occurring, but it is important to correctly identify the specific impacts the gambling activities can have on a person.

The Addiction

The addictive nature of gambling is related to neurotransmitters, such as dopamine, within the body. Whenever someone is engaged in enjoyable activities such as eating, sex, drugs, and situations where the reward is uncertain, the brain releases dopamine (Robinson, 2018). High amounts of dopamine lead to immense levels of pleasure within a person, which is why some researchers believe that gambling is an activity that many use as a replacement for the satisfaction they lack in their own life (Balodis & Potenza, 2020). Additionally, some parts of the brain such as the thalamus are related to addiction caused by gambling due to the concept of loss aversion. Loss aversion describes the tendency of a person to choose gains and avoid losses, and it is related to norepinephrine levels (Takahashi et al., 2013). Studies have shown heightened levels of norepinephrine in pathological gamblers, which directly impact the neurotransmitters' binding potential within the thalamus, consequently leading to decreased loss aversion parameters (Balodis & Potenza, 2020; Takahashi et al., 2013). Fewer loss aversion parameters within a person suggests that they would become immune to losing money because they would see nothing wrong with it, hence continuing the same action of gambling without second thought. Although a gambling addiction is typically developed over time, genetic factors have also been linked to these disorders. A 2011 study by Slutske et al. concluded that genetic factors contributed to 49.2% of disordered gambling, as defined by one overseeing body, and

54.4% of disordered gambling as per another (Slutske et al., 2011). These genetic factors are related to variants on specific genes that make an individual more susceptible to gambling. Regardless of how one develops a dependence on gambling, it is important to address the issue before it leads to adverse long-term effects

Changes Caused by the Pandemic

With the start of the pandemic, many families and individuals were faced with financial uncertainty, which might have led to a phenomenon known as problem gambling. Problem gambling, brought forward by financial and mental health problems, occurred before the pandemic in areas all around the world, and has even happened during previous national and international financial crises (Håkansson et al., 2020). Prior to the past year, online gambling had been gaining traction due to its availability and ease, which can make it more problematic compared to traditional gambling markets. Online gambling becomes a health hazard when it is fueled by individuals that are at home dealing with anxiety, psychosocial stress, and depression (Håkansson et al., 2020). Of a group of 2005 gamblers surveyed, 54% of them had gambled online during the emergency measures in Ontario, Canada (Price, 2020). It was seen that play of card games and slot games appeared to have increased in May, a few months into the lockdowns, possibly due to the shift to online modes of play (Turner, 2020). The numbers could be attributed to the closure of many in-person gambling venues, but there was still a strong gambling presence amongst the community. A few months into the lockdowns seemed to motivate more people to make the switch to online gambling. The same survey noticed that males were more likely to gamble online compared to females, and it was more common amongst younger adults aged 18-24 compared to those older (Price, 2020). The surge in younger gamblers can be connected to the fact that many young individuals had jobs in the service sector, many which were shut down during the pandemic. These people might have considered gambling a platform through which they can mitigate the financial uncertainty.

Elsewhere, research done by Auer and Griffiths showed that the amount of active players for gambling platforms increased steadily from January 1, 2020 to May 31, 2020, and the average daily bet on May 31 was 9% higher than that on January 1 (Auer & Griffiths, 2021). Similar to trends in Canada, many other confinement orders in countries around the world led to a collective towards online better. Another study found that although online casino gambling did not specifically become more frequent, those who frequently bet on sports continued to bet (Auer & Griffiths, 2021). Not only that a decent amount of participants in a secondary survey shared how the lack of options for sports betting led them gamble on other sporting events more than they typically would (Håkansson, 2020). These statistics are reflective of the impact that these addictions have on a person, because regardless of the situation, an individual will always find a way to get their dose of satisfaction. The number of high-risk players also decreased during these months, yet the total amount players have increased through new registrations, potentially because of heightened financial uncertainty.

Mental Health Impacts

It is evident that the pandemic brought on an onset of increased gambling, but that gambling also brought forward some deeper mental health concerns within the general population. Through a survey regarding the first few weeks of the COVID-19 lockdowns, it was found that in those seeking treatment for gambling disorders, many were worried about the work and the risk of infection (Håkansson et al., 2020). Constant worries about finances and health leave many people vulnerable to opportunities they would not traditionally consider. After two weeks of confinement, 12% reported worse gambling, 19% were abstinent, 46% showed signs of anxiety, and 27% showed signs of depression (Håkansson et al., 2020). At another checkpoint a while later, it was revealed that about 57% of respondents had anxiety issues ranging from mild to severe, and 56% had symptoms of depression ranging mild to severe (Price, 2020). In both situations, most of those reporting anxiety or depression were found to have mild symptoms, with only a small percentage reporting severe symptoms. The data suggests a direct correlation between the lockdowns, mental health, and gambling as they all go hand in hand to negatively impact vast populations. The rise in the anxiety and depression symptoms is concerning because during the pandemic, it has been difficult for many to get the support they need to combat any illness. Additionally, a third of gamblers surveyed said how they had experienced some sort of mental health issue before the pandemic, which seemed to only get quantified by the hardships they had to face once the lockdowns started. A major concern associated with these increasing mental health problems is the substance abuse that often follows and was seen in the gamblers. Majority of those surveyed reported consuming alcohol during the first six weeks, with 40% saying that their consumption had increased, while about 50% reported increased cannabis use (Price, 2020). Although not all alcohol and cannabis use are a health hazard, it brings forth the possibility of other comorbidities along with the gambling addiction. Both seem interconnected through the mechanisms of impulse control issues, which lead to high-risk taking behaviour seen in both substance use and gambling (Ford & Håkansson, 2020).

Conclusion

Overall, it is unmistakable that the COVID-19 pandemic has brought hardships in various forms on many people. The constant lockdowns, financial instability, and major mental health concerns have made many individuals vulnerable to gambling. Those that were already struggling with a gambling habit or addiction have found themselves struggling again, whereas new members are susceptible to developing an addiction. The gambling industry is beneficial for many economies around the world, yet gambling can have severe complications on an individual, many which are long lasting. The increasing symptoms of anxiety and depression amongst gamblers, as well as rising usage of alcohol and marijuana portrays the need to address the situation. The long-term effects of the pandemic on the world are unknown, which is why it is crucial to monitor the situation in the gambling industry to ensure that populations are being safe and responsible.

References

Auer, M., & Griffiths, M. D. (2021). Gambling Before and During the COVID-19 Pandemic Among Online Casino Gamblers: An Empirical Study Using Behavioral Tracking Data. International Journal of Mental Health and Addiction. https://doi.org/10.1007/s11469-020-00462-2

Balodis, I. M., & Potenza, M. N. (2020). The Biology and Treatment of Gambling Disorder- ClinicalKey. https://www-clinicalkey-com.libaccess.lib.mcmaster.ca/#!/content/book/3-s2.0-B9780323754408000337

Ford, M., & Håkansson, A. (2020). Problem gambling, associations with comorbid health conditions, substance use, and behavioural addictions: Opportunities for pathways to treatment. PLOS ONE, 15(1), e0227644. https://doi.org/10.1371/journal.pone.0227644

Håkansson, A. (2020). Impact of COVID-19 on Online Gambling – A General Population Survey During the Pandemic. Frontiers in Psychology, 11. https://doi.org/10.3389/fpsyg.2020.568543

Håkansson, A., Fernández-Aranda, F., Menchón, J. M., Potenza, M. N., & Jiménez-Murcia, S. (2020). Gambling During the COVID-19 Crisis – A Cause for Concern. Journal of Addiction Medicine, 14(4), e10. https://doi.org/10.1097/ADM.0000000000000690

Price, A. (2020). Online Gambling in the Midst of COVID-19: A Nexus of Mental Health Concerns, Substance Use and Financial Stress. International Journal of Mental Health and Addiction, 1–18. https://doi.org/10.1007/s11469-020-00366-1

Robinson, M. (2018). Designed to deceive: How gambling distorts reality and hooks your brain. The Conversation. http://theconversation.com/designed-to-deceive-how-gambling-distorts-reality-and-hooks-your-brain-91052

Slutske, W. S., Zhu, G., Meier, M. H., & Martin, N. G. (2011). Disordered Gambling as Defined by the DSM-IV and the South Oaks Gambling Screen: Evidence for a Common Etiologic Structure. Journal of Abnormal Psychology, 120(3), 743–751. https://doi.org/10.1037/a0022879

Takahashi, H., Fujie, S., Camerer, C., Arakawa, R., Takano, H., Kodaka,

F., Matsui, H., Ideno, T., Okubo, S., Takemura, K., Yamada, M., Eguchi, Y., Murai, T., Okubo, Y., Kato, M., Ito, H., & Suhara, T. (2013). Norepinephrine in the brain is associated with aversion to financial loss. Molecular Psychiatry, 18(1), 3–4. https://doi.org/10.1038/mp.2012.7

Turner, N. E. (2020). COVID-19 and Gambling in Ontario. Journal of Gambling Issues, 44(0), Article 0. https://doi.org/10.4309/jgi.2020.44.1

PANEL 3: HISTORICAL REFLECTIONS

The *Erebus* Wreck:
Scientific Discoveries and Remaining Questions

Christina MacDonald

McMaster University

Hamilton, Ontario, Canada

Author Note

Research on this topic was originally completed for the Antarctic Institute of Canada's Level Up Program. These findings were originally written for the book The Wreck of the Erebus and have been adapted for this paper. The information in this paper was also adapted for a presentation at the Antarctic Institute of Canada's Second Annual Conference. There are no conflicts of interest to be disclosed.

Correspondence concerning this paper should be addressed to Christina MacDonald, McMaster University. Email: macdoc21@mcmaster.ca

Abstract

The recently discovered wreck of the HMS *Erebus*, which set sail over 170 years ago, provides an insight into history. Using the current studies of bioarchaeology, forensics and chemical analysis, scientists have been able to study recovered artifacts and remains to better understand the nature of the wreck and the lives lost on it. The scientific fields work together to collect, analyze, and interpret data from biological samples. This has led to successful identification of remains as well as given insight into different conditions the bodies endured. There have been great advances, however, there are still many unanswered questions regarding the *Erebus* wreck. Future expeditions and further research must be conducted to locate and analyze artifacts, better understand why the ship sank, and uncover what other lives were lost.

Keywords: bioarchaeology, forensic science, chemical analysis

The Basics About HMS *Erebus*

HMS *Erebus* was a British ship manned by over 100 men and led by an experienced explorer named Sir John Franklin (Mulvaney, 2020). HMS *Erebus* set sail with another ship, HMS Terror, but this paper only focuses on HMS *Erebus*. The expedition's goal was to locate the Northwest Passage, a connection between the Atlantic and Pacific, through the Arctic (Mulvaney, 2020). The importance of this passage is that it would facilitate England's travel and trade. What made this such a treacherous expedition was the icy waters and difficult weather conditions in the Arctic. Even after the ships were equipped to deal with harsh conditions the ice proved to be mightier than the vessels (Mulvaney, 2020). HMS *Erebus* set sail 176 years ago in 1845, but the wreck was only discovered in 2014 by Parks Canada with the help of Inuit oral traditions (Mulvaney, 2020). There were oral recounts of Inuit witnessing crew members fleeing HMS *Erebus* and of where this ship sank, but these were disregarded for a long time. Although it took over a century, since the 2014 discovery many Parks Canada expeditions have collected artifacts and data to try and piece together the great mystery of the *Erebus* wreck. As society continues to expand scientific knowledge in the future, future advances will reveal more about the events of the past.

There are a plethora of scientific methods currently being used to aid in uncovering details about the *Erebus* wreck. Some notable scientific methods that will be further explored in this paper are bioarchaeology, forensic science and chemical analysis. Moreover, this paper will delve into some of the unanswered questions about the wreck.

The Basics of Bioarchaeology

Archaeology is a term that many people may be familiar with, bringing to mind images of the pop culture depictions of archaeological digs in movies or memories of history lessons about the discoveries of mummies and artifacts. What may be lesser known are the subcategories of archaeological study. In general, archaeology is the study of remains, but it can be further divided into categories such as bioarchaeology—the study of human remains (Society for American Archaeology, n.d.).

Bioarchaeology is a field that works towards recovering the lives of those that lived and that were lost hundreds, or even thousands of years ago, but a unique challenge facing the archaeological expeditions in relation to the *Erebus* wreck is that complications arise when an archaeological site is underwater. In the freezing Arctic waters, there are many restrictions and divers require intense training and must be equipped with special gear to efficiently dive and document findings (Parks Canada, 2019). Being in a remote area, the logistics behind such archaeological expeditions are no small feat—preparations must be done well in advance, with planes and boats being the main modes of transportation for many of the needed resources (Parks Canada, 2019).

The Basics of Forensics

Moving away from bioarchaeology, forensic science may also be linked to pop culture as it is often associated with crime scene investigation or crime shows that help lock up perpetrators through DNA analysis— the analysis of samples of biological materials (Canadian Society of Forensic Science, n.d.). Although these depictions of the field are often sensationalized, forensic science provides great insight into the lives of the deceased. Forensics utilises scientific disciplines such as biology, physics, chemistry, hair and fibre analysis, as well as other disciplines such as odontology—the study of dental evidence (Canadian Society of Forensic Science, n.d.). One field that will be further explored as it has proven to be very useful in studying the *Erebus* wreck is DNA analysis.

DNA analysis is an area within forensics that is integral to the study of remains as it gives indications to the identity of otherwise nameless skeletons. Even after remains have been exposed to the elements for significant periods of time, bones and teeth still maintain some extractable information because their hard structures protect against DNA breakdown, or DNA degradation (Latham and Miller, 2018). DNA analysis can also be used to examine hair samples that have maintained some DNA. One factor that is often taken into consideration in the process of DNA degradation is temperature; warmer temperatures being more detrimental to the salvaging of DNA (Latham and Miller, 2018). The cold climates of the Arctic, the place in which the remains of the *Erebus* wreck are located, likely plays a role in the maintenance of forensic evidence.

Any collected samples are handled very carefully to conserve the maximum amount of extractable DNA (Latham and Miller, 2018). When conducting the multi-step process of preparing and testing samples, scientists must take into consideration that DNA analysis is a complex and delicate task so environmental and chemical factors need to be controlled in the lab. All work areas must be sterile to prevent contamination and all workers must always enter the work area in personal protective equipment—this includes disposable gloves and goggles (Edson and McMahon, 2016).

Another important factor to consider is the type of DNA that can be extracted from human remains. Human cells have nuclear and mitochondrial DNA. Nuclear DNA is located in the cell's nucleus—the part of human cells that contain structures called chromosomes which carry the genetic code that makes every individual unique (Latham and Miller, 2018). The standard number of chromosomes is 46; each cell nucleus in the body will only have one copy of the 46 chromosomes (Latham and Miller, 2018). Nuclear DNA can give great insight into the identity of the remains because it can be linked to a person's entire ancestry. Unfortunately, this type of DNA is not always attainable. Mitochondrial DNA, on the other hand, is found in a different part of the cell—the mitochondria. The mitochondria is another structure within human cells but the genetic information it contains is more resistant to degradation than that in the nucleus because it contains several copies of genetic information (Latham and Miller, 2018). A big difference between the types of DNA is that mitochondrial DNA only provides information about maternal DNA, meaning that to be able to make any identifications, scientists can only look at the genetics of the biological mother's lineage (Latham and Miller, 2018). However, just as mitochondrial DNA gives insight into the biological mother's gene line, analysis of available Y-chromosomes in nuclear DNA allows for the tracking of the biological father's lineage. Even with all the information DNA can provide—notably, it can help with identification and the tracing of remains to living relatives—chromosomes do not determine everything about a person's life and identity.

A couple of other useful areas of forensics are fingerprinting and facial reconstruction. By analyzing the unique patterns of finger ridges left behind on surfaces, forensic scientists, in collaboration with other professionals such as police officers, can sometimes match the found prints with fingerprints recorded in databases (National Institute of Justice, 2013). In relation to the *Erebus* wreck, a sole fingerprint was found in a wax seal. There may not be much significance to this finding for forensic tests right now, but it still gives indication of a life lost. With advancements in technology, a simple fingerprint may one day help uncover another crucial detail about the otherwise unknown lives of the past. Stepping away from the science of the discovery, fingerprints also serve as a reminder that the discovered bones once belonged to a person filled with life. In regards to facial reconstruction, artists and forensic anthropologists—who study human remains—can use clay to model the face that may have accompanied a skull (Gupta et al., 2015). Using specific techniques and by looking at various indicators, skin thickness and general facial features can be estimated (Gupta et al., 2015). With all the information forensics can reveal, it must be acknowledged that it would not be as fruitful if it were not for chemical analysis.

The Basics of Chemical Analysis

The basic purpose of chemical analysis is to identify the composition of various samples being tested in a lab (Braun, 2016). This process allows scientists to determine the physical properties of different materials. In the bioarchaeological and forensic contexts of analyzing bones, chemical analysis provides information about lifestyle, nutrition, and where a person was from (Jansen, 2018).

Various elements—the simplest forms of substances which cannot be broken down any further—allow for conclusions to be drawn (Myers, 2012). In the body, there are different forms of elements like carbon, nitrogen, and oxygen—in chemistry, these variations are called isotopes (Herzog, 2021). For example, when analyzing bone and teeth samples, depending on what form of carbon is found, or in other words, what carbon isotope is found, one may be able to determine what types of foods were eaten by the deceased person (Jansen, 2018). Such trends in food consumption can then be linked to different areas in the world, providing further information about the remains.

Putting it All Together

Even though these are separate fields, when studying the Erebus wreck one field of study cannot fully function without the other. Bioarchaeologists cannot fully understand remains without DNA analysis; forensic scientists cannot fully understand the results of their tests without the use of chemical analysis. To use an Arctic-inspired analogy, bioarchaeology is only the tip of the iceberg. It is an active process that yields tangible artifacts and findings, but that is simply the surface. Forensic science and chemical analysis then come into play when looking at the details hidden beneath the surface and often reveal the more critical information that would have otherwise gone unnoticed. When looking deeper into the discoveries of the *Erebus* wreck, all scientific methods work together to tell a more complete story.

There have been multiple remains found from the wreck over the past several decades—a sizable number of deaths occurring on Nunavut's King William Island—but only more recent advances have led to their identification. A bioarchaeological sample, in this case, a skull was found at King William Island and scientists were able to recover DNA from it. After locating living descendants and acquiring genetic evidence from Y-chromosome testing, scientists were able to identify the skull as John Gregory, an engineer from the HMS *Erebus* (Potter, 2021; Stenton et al., 2021).

Proceeding identification, facial reconstruction was conducted on John Gregory's skull, bringing a face to the archaeological discovery. Facial reconstructions have been used for various other discovered remains and may even indicate that some of the initial assumptions of the identity of the skulls were incorrect (Stenton et al., 2015). Even though facial reconstruction is not biological evidence, it is able to help strengthen (or question) what is currently known about the discovered bodies from the wreck.

Living descendants play a significant role in being able to confirm the scientific evidence revealed by bioarchaeologists, forensic scientists and chemical analysts. They help to bridge the gap between the scientific information extracted from bones that are hundreds of years old to people living today. The identification of John Gregory was a well-reported one, but other skeletal remains have also yielded genetic and chemical information that has been helpful in determining the approximate age, height, and health status of the deceased crewmen (Gershon, 2021).

Although new advances will shed light on a lot of interesting aspects of the events leading up to and following the *Erebus* wreck, they may also reveal some disturbing truths. Moving away from the genetic aspect of the findings, bones can be analyzed for different types of markings that indicate what the bodies may have endured before and after death. A more morbid conclusion drawn from research looking at different breaks in bones is that there were likely acts of cannibalism amongst the crewmembers; this is also backed up by Inuit oral tradition (Mays and Beattie, 2016; Mulvaney, 2020). From connecting the lives lost in the wreck to living relatives to uncovering more about what the crewmembers endured in their final days, there are many more discoveries to be made. However, just as more conclusions are drawn, more questions continue to arise.

Unanswered Questions

There is still much that is unknown about the nature of what happened to the individuals that made the fateful journey on the HMS *Erebus*. Current discoveries in identification have been able to help solve many mysteries, but they also draw many questions.

Who were the people on board at the time of abandonment? What will further DNA testing uncover? There are still several unidentified bones, and even more that remain unfound, so it is intriguing to think about how much more may be learned with further discoveries and as technology and identification practices continue to advance. This also brings up the question: are the current identifications truly accurate? For example, there is some speculation about whether the current identification of Lieutenant John Irving's remains is accurate, or if the remains are of a different officer (Warrior, 2020). There have been rectified mistakes in the past—some remains were originally identified as Lieutenant Henry Le Vesconte, but then were more recently corrected to be an assistant surgeon by the name of Harry Goodsir (Warrior, 2020). This goes to show that the process of identification is not perfect, in order to seek out the truth, it must be scrutinized and actively questioned. There are also individual factors that must be considered in regards to locating and identifying remains. For example, larger bones are more resistant to degradation and have a better rate of survival (Latham and Miller, 2018). The unknown rate of degradation of the bodies and the different areas in which they may be found makes it very difficult to predict what will be discovered in the future.

Will artifacts found shed more light on the events of the past? With every expedition, more is uncovered and understood. The fingerprint in the wax seal that was discussed earlier may not seem like the best piece of forensic evidence to bring names to the lost faces, but it may lead to larger discoveries with time. All the artifacts have the potential to tell a greater story, acting as individual pieces in the great puzzle that is the *Erebus* wreck.

How did COVID-19 impact discoveries? The global pandemic was an unexpected barrier in recent expeditions. Regarding the wreck expeditions, COVID-19 impacted the ability of teams to explore and gather data. Ensuring the safety of the vulnerable and high-risk communities in Nunavut took precedence over exploration (CanadaNewsWire, 2020). The possible discoveries 2020 could have brought will remain a mystery, but advancements in knowledge and planning for future expeditions will persist. In an article approved by Parks Canada, it was revealed that more efforts would be put into research into the already-recovered items (Cision, 2020). Even with unexpected challenges that bring up new questions on how to approach the wreck and its artifacts, there is still great potential for advancement.

It took close to 200 years for the wreck to be found; the technology needed to find it was not available at the time of the wreck. Even with the great technology of the 21st century, there may still not be resources sophisticated enough to uncover all the unknowns about the fate of the HMS *Erebus*, leaving many remaining questions. How did the HMS *Erebus* sink? What lives were tragically lost? What else is on the *Erebus* wreck? There are so many questions that can be asked, but arguably the most exciting question of all is: What will be uncovered next?

References

Braun, R. Denton (2016). Chemical analysis. Encyclopedia Britannica. https://www.britannica.com/science/chemical-analysis

CanadaNewsWire (2020). Parks Canada shifts plans for 2020 archaeological fieldwork at the wrecks of HMS Erebus and HMS Terror in response to COVID-19. http://libaccess.mcmaster.ca.libaccess. lib.mcmaster.ca/login?url=https://www-proquest-com.libaccess.lib. mcmaster.ca/wire-feeds/parks-canada-shifts-plans-2020-archaeological/ docview/2417474219/se-2?accountid=12347

Canadian Society of Forensic Science. (n.d.). What is forensic science. https://www.csfs.ca/student-zone/student-zone/

Cision. (2020). Parks Canada shifts plans for 2020 archaeological fieldwork at the Wrecks of HMS Erebus and HMS Terror in response to COVID-19. https://www.newswire.ca/news-releases/parks-canada-shifts-plans-for-2020-archaeological-fieldwork-at-the-wrecks-of-hms-erebus-and-hms-terror-in-response-to-covid-19-825618055.html

Edson, S. M., & McMahon, T. P. (2016). Extraction of DNA from Skeletal Remains. Methods Mol Biol, 1420, 69–87. https://doi.org/10.1007/978-1-4939-3597-0_6

Gershon, L. (2021). Descendant's DNA helps identify remains of doomed Franklin expedition engineer. Smithsonian Magazine. https://www.smithsonianmag.com/smart-news/doomed-arctic-explorer-idd-dna-180977678/

Gupta, S., Gupta, V., Vij, H., Vij, R., & Tyagi, N. (2015). Forensic Facial Reconstruction: The Final Frontier. J Clin Diagn Res, 9(9), ZE26–ZE28. https://doi.org/10.7860/JCDR/2015/14621.6568

Herzog, G. F. (2021). Isotope. Encyclopedia Britannica. https://www.britannica.com/science/isotope

Mulvaney, K. (2020). What happened to the doomed Franklin expedition? These are the clues. HISTORY. https://www.history.com/news/franklin-expedition-mystery-northwest-passage

Jansen, K. (2018). Chemical analysis of bones fills gaps in history. Chemical and Engineering News, 96(21).https://cen.acs.org/analytical-chemistry/art-&-artifacts/Chemical-analysis-bones-fills-gaps/96/i21

Latham, K. E., & Miller, J. J. (2018). DNA recovery and analysis from skeletal material in modern forensic contexts. Forensic Sciences Research, 4(1), 51–59. https://doi.org/10.1080/20961790.2018.1515594

Mays, S., & Beattie, O. (2016). Evidence for end-stage cannibalism on Sir John Franklin's last expedition to the Arctic, 1845. International Journal of Osteoarchaeology, 26(5), 778-786. https://doi.org/10.1002/oa.2479

National Institute of Justice. (2013). Fingerprints: An overview. U.S. Department of Justice. https://nij.ojp.gov/topics/articles/fingerprints-overview

Parks Canada. (2019). Wrecks of HMS Erebus and HMS Terror National Historic Site. Government of Canada.https://www.pc.gc.ca/en/lhn-nhs/nu/epaveswrecks/culture/archeologie-archeology/explore/subaquatique-underwater

Potter, R. (2021). 1845 Franklin expedition member identified using DNA. Canadian Geographic. https://www.canadiangeographic.ca/article/1845-franklin-expedition-member-identified-using-dna

Society for American Archaeology. (n.d.) What is archaeology? https://www.saa.org/about-archaeology/what-is-archaeology

Stenton, D. R., Fratpietro, S., Keenleyside, A., & Park, R. W. (2021). DNA identification of a sailor from the 1845 Franklin northwest passage expedition. Polar Record, 57, E14. doi:10.1017/S0032247421000061

Stenton, D. R., Keenleyside, A., Trepkov, D. P., & Park, R. W. (2015). Faces from the Franklin expedition? Craniofacial reconstructions of two members of the 1845 northwest passage expedition. Polar Record, 52(1), 76-81. https://doi.org/10.1017/S0032247415000248

Warrior, C. (2020). New discoveries from the lost Franklin expedition. Royal Museums Greenwich. https://www.rmg.co.uk/stories/blog/curatorial/new-discoveries-wrecks-hms-erebus-terror

The Meanders: A Geographical Contextualization of Heraclitus' Ever Changing River Fragments

Jilene Malbeuf[1,2], A. A. Mardon[2], Z. Schauer[2,3], and J. Jutras[2,3]

[1]University of Alberta, Edmonton, Canada
[2]Antarctic Institute of Canada, Edmonton, Canada
[3]MacEwan University, Edmonton, Canada

Author Note

Correspondence concerning this article should be addressed to Jilene Malbeuf, email: jilene@ualberta.ca.

Abstract

Heraclitus is best known for his fragmentary statement, "you cannot step into the same river twice." He is primarily concerned with the constantly changing contents of a river as the water flows downstream, but the nature of the rivers that he would have been familiar with are often overlooked. His home in Ephesus is situated near two rivers which are classified as meandering streams, meaning their banks and course also change over time. This added depth to his statement provides a context to Heraclitus that is often missing, and it provides an argument for the merits of interdisciplinary studies and the value of localized knowledge.

Keywords: Heraclitus, Ephesus, meandering streams, fragments, river, change, philosophy, geography, setting, context, interdisciplinary.

Heraclitus of Ephesus is a Pre-Socratic philosopher well known for his short thought-provoking fragments conveniently formatted to provide starting points for stimulating discussion. One of his most popular starting points talks of never being able to step into the same river twice. It is clear in the relevant fragments that he is primarily referring to how the water is constantly changing as it flows downstream, arguably making the river new and different every moment as its contents change. An understanding of the setting in which Heraclitus mused and wrote can contextualize his thoughts and provide a separate meaning beyond this that is not often considered. His location on the Aegean coast places him near rivers that change more than just their contents, as their routes also change over time. This little detail would likely have been apparent to locals as well as his contemporaries, but it has been lost as his work has been analyzed globally. The thoughts of Heraclitus may have been reduced to a series of fragmentary thoughts over time, but that does not mean that they exist without any context to add depth to their meaning.

Heraclitus was active around 500 BCE in Ephesus, which was a prominent port city in Asia Minor (near modern day Kuşadası, Turkey) (Graham, 2021). Little is known about his life, as the book that he is known to have written and dedicated to the temple of Artemis only exists today through quotations from other later authors. Although he is kept separate from the nearby Milesian philosophers, his ideas suggest that he was aware of them. One of his theories is that all is fire, which follows the Milesian tradition of searching for a rational order to the universe that can be reduced to a single form of matter. His other major theories include the postulations that opposites coincide and that things are in a constant state of change (Graham, 2021).

This last theory can be seen in his discussion of rivers, which is extant in three separate fragments. Of the three, fragment 91 from Plutarch has been shown to be closest to what would have been his precise words in his book: "You cannot step twice into the same rivers; for fresh waters are ever flowing in upon you." As Graham explains, the statement is, on the surface, paradoxical, but there is no reason to take it as false or

---

[1] Thales argued that all is water, Anaximander claims that all is apeiron (infinite, indefinite), and Anaximenes says that all is air.

[2] Fragment 12 reads "Upon those who step into the same rivers different and different waters flow," and fragment 49a reads "We step and do not step into the same rivers; we are and are not." For a discussion of why fragment 91 is preferred, see Graham, 2021.

contradictory. It makes perfectly good sense: we call a body of water a river precisely because it consists of changing waters; if the waters should cease to flow it would not be a river, but a lake or a dry streambed (Graham, 2021). This idea not only challenges the natural way of understanding rivers, but also what constitutes sameness and difference. The river can simultaneously be the same and different, providing an example of his conception of the unity of opposites. Thus, this fragment (or series of fragments) constitutes an entertaining thought experiment that lends itself nicely to many an Intro to Philosophy course, and so is usually the way in which people learn about Heraclitus.

This customary direction that discussions of Heraclitus' river fragments take may be fulfilling in itself, but it may not cover the entirety of his original intention with these musings. The water within a river may be constantly changing, but this is not the only way that rivers can change. Some rivers are classified geographically as meandering streams, which wind through a landscape and through erosion and sedimentation change their banks over time. As the river twists and turns, the maximum velocity of the current is knocked out of center, eventually hitting the outer bank of any bends. This causes the moving water to eat away at the outer bank and sediment to be deposited on the calmer inner bank, in time moving the river sideways and slightly downstream. In extreme cases, this process can cause the beginning and end of a bend to meet, causing the river to straighten out to follow the most direct course, eventually cutting off the bend to create a separate oxbow lake (U.S. National Park Services, 2020).

The phenomenon of meandering rivers is something that Heraclitus would very likely have been aware of. The Milesian philosophers who were his contemporaries could be found nearby, as Ephesus and Miletus are both in the region of Ionia (Encyclopedia Britannica, 2012).[3] There is no record of direct contact between these thinkers and it is not likely that Heraclitus ever traveled to Miletus, but it is clear in his fragments that he was aware of this other city and its philosophers, as some of his theories seem to respond to them in one way or another (Graham, 2021). Between these two cities lies one of the major landmarks of the region: the Maeander River (Encyclopedia Britannica, 2012).[4] The English word "meander" along with the geographical term described above are both derived from this particular river. As the inspiration for the term, it demonstrates all of the

---

[3] These cities are roughly 100km apart.

[4] *Büyük Menderes Irmağı* in modern Turkish.

trademarks of a meandering river, which Heraclitus would have been a witness to. At its closest point today, the Maeander River is approximately 30km from the archaeological site of Ephesus, so while its banks have changed in the intervening years it would have still been close enough for Heraclitus to have been aware of.

There is a river even closer to Ephesus that also shares these qualities. The Caystrus River is another major river within the region of Ionia, which today bears the name Küçük Menderes or "Little Meander" (Encyclopedia Britannica, 2012). In Heraclitus' time, Ephesus was a port city where this river entered the Aegean Sea, in the same way that Miletus was a port city on the Maeander. Today, however, both sites are several kilometers inland, as through time sedimentation from these rivers has filled in their corresponding inlets. These major impacts on the landscape would have been apparent in 500 BCE as well as today, and so would have been integral to how Heraclitus conceived of rivers. Thus, while he was primarily concerned with the changing waters within rivers in his discussion of them, he would have also been aware that riverbanks are not fixed and unchanging aspects of rivers either.

In conclusion, there is more than one reason why one cannot step into the same river twice according to Heraclitus. Not only do the contents change, but so too can its banks and route over time. This additional geographical knowledge does not provide an additional interpretation to his river fragments; instead, it works to reinforce how it has always been understood. His overarching theories that things are often in a constant state of change and that opposites need not be mutually exclusive can be felt in both phenomena. Change is shown to be an unavoidable aspect of a river's existence, and so constancy and change are inextricably linked in its existence. The characteristics of meandering rivers may have been well known to Heraclitus' interlocutors, but this knowledge is not so common today. Becoming aware of this circumstantial knowledge adds contextual depth to the thoughts of an ancient philosopher who is notorious for being otherwise entirely without context. With this in mind, even if the intentions of this fragment remain the same this added knowledge is valuable.

References

Encyclopedia Britannica. (2012, March 6.) Ionia. https://www.britannica.com/place/Ionia.

Graham, Daniel W. (2021, Summer.) Heraclitus. The Stanford Encyclopedia of Philosophy. https://plato.stanford.edu/archives/sum2021/entries/heraclitus/.

U.S. National Park Services. (2020, April 22.) Fluvial Features - Meandering Stream. https://www.nps.gov/articles/meandering-stream.htm.

PANEL 4: CONTEMPORARY POLITICAL ISSUES

Democracy and Legitimacy in the Canadian Senate

Benjamin A. Turner and Dr. Austin A. Mardon, CM, KCSS, FRSC, FRCGS

Abstract

The structure of the Canadian Senate has been a consistent political tripping hazard for those in power since its addition to the governance structure of Upper and Lower Canada in 1791. Changes in the structure and function of the Senate were numerous in the nineteenth century, with varied levels of success, and while the Senate has remained largely unchanged through the 20th century, there have been many attempts to change the structure and function since 1980, with the most consistent and substantial change being to the appointment process. This paper will discuss whether changing to a democratic system for selecting Senators would lend more legitimacy to the Upper Chamber, and what the potential consequences would be for the function of the legislative branch of the Canadian government.

Keywords: Senate, democracy, legitimacy, Canadian government, Senate reform, Supreme Court, constitutional amendment, politics, Canadian politics

Introduction

The question of legitimacy in the Senate is as old as the upper chamber itself. The first form of the upper chamber in 1791 was hereditary (O'Brien, 2019), this state of affairs triggered the first Senate reform in relatively short order, by 1810 Canada had an appointed but no longer hereditary upper chamber. This was far from the last difficulty the Senate experienced. The quest for greater legitimacy in the Canadian Senate appears to be one of our oldest pastimes, but what exactly does a legitimate Senate look like? All the Senate reform proposals since 1980 have had a common factor: democracy. Each commission and proposal proposed a different balance for how to reform the Senate to be more effective, more regionally balanced, and how to shake up the legislative powers of parliament by emphasizing the Senate in some matters while leaving most legislative responsibilities in the hands of the House of Commons, but they all identified democracy as an important part of the solution.

In this paper, I will argue that yes, perhaps a democratically elected Senate may enjoy more legitimacy with the political body in superficial terms, but this solution affects more than just the appointment process of Senators. When considering legitimacy in terms of both the appointment process and the legislative function of the Senate within government, I find that electing Senators grants legitimacy while at the same time potentially undermining its function as a complementary body in the bicameral model of Canadian federalism (Hulme, 2016). Simply put, the answer to the question of whether a democratically elected Senate would be more legitimate is: yes and maybe not.

Defining Legitimacy

There have been many major attempts at Senate reform in Canada in the last 40 years, including the Charlottetown and Meech Lake accords, and the 2014 Supreme Court Senate Reference case. None so far have been implemented, and all placed the democratic selection of Senators as the primary solution, but would a democratically elected Senate have fixed the underlying issues of the perceived lack of confidence in the upper chamber? To determine the potential effectiveness of these reform proposals, it is necessary to replace our implicit definition of legitimacy with an explicit one. I have reviewed the work Stillman (1974) to fill in this blank. Stillman puts it this way: "A government is legitimate if and only if the results of governmental output are compatible with the value pattern of the society." In applying this definition to Senate reform, government output is prudently viewed as both the appointment process for Senators, as well as how well the Senate would likely be to perform its function in government as a chamber of sober second thought. The crucial element of this function within Canadian federalism is that the Senate is not to obstruct or undermine the House of Commons, it is to offer constructive suggestions to improve legislation and then send it back to the elected lower house for final reading.

Reform Attempts

There are eight major reform attempts and committee recommendations that I have reviewed, including the more recent changes made by Prime Minister Justin Trudeau. The changes made by Mr. Trudeau are unique for a few reasons, notably: he does not call for democracy in the Senate. He dissolved the liberal caucus while still in opposition, and when in government established an independent advisory board for the appointment of new Senators, granting liberal appointees a level of autonomy not seen in the Senate since 1848-1867 (O'Brien, 2019). These changes show respect for the nature of the Senate as a complementary body of sober second thought (Hulme, 2016). And perhaps most importantly, since the Trudeau changes are not subject to the near impossible standard of a constitutional amendment (Docherty, 2002), they have been successfully implemented.

In the autumn of 2013, the government under Prime Minister Stephen Harper submitted a set of reference questions regarding Senate reform to the Supreme Court (Hulme, 2016). The legislation Mr. Harper was testing would require every province in Canada to enact Senator-in-waiting legislation similar to the one on the books in Alberta at the time. The function of this legislation was to put Senate candidates on the ballot any time there was a vacancy for the province in question; voters of that province would elect a candidate and the Prime Minister would subsequently appoint that individual. Stephen Harper referred this proposed legislation to the Supreme Court because he aimed to use parliament to unilaterally alter the Senate, thus side-stepping the issue of passing a constitutional amendment which was the roadblock that sank the Charlottetown and Meech Lake Accords (Docherty, 2002). The Supreme Court, as previously mentioned, blocked this attempt to bypass constitutional amendment and offered a specific and narrow definition for the function of the Senate, primarily focusing on its complementary function and eliminating considerations like regionality and the balance of legislative power in Parliament, reinforcing the supremacy of the House of Commons in such matters.

This judgement effectively killed Prime Minister Harper's attempt at Senate reform. After years of claiming that he could accomplish reform without the need for messy constitutional negotiations, the Supreme Court solidly shut the door on the possibility (Hulme, 2016).

This interpretation of the function of the Senate speaks to the matter of legitimacy because it lends absolute clarity to the part the Senate is intended to play in Canadian governance and firmly establishes the need for a constitutional amendment to alter that role. On the topic of output, if the Senate is to be a complementary body to the House of Commons, it is likely that elected Senators would feel emboldened if they were selected directly by voters. This would entitle them to the responsibility of interpreting the will of voters and introducing their own original legislation, and give them a mandate to obstruct the House of Commons in the legislative process. Having a mandate to govern and interpreting the will of the political body is not a bad thing, but that responsibility already rests in the hands of the House of Commons. An elected Senate, therefore, would be not just a duplication of efforts, but if the balance of power is different between the upper and lower chambers, the Senate is entirely likely to become an obstructive body. The result would be a Senate that undermines the work of elected Members of Parliament in the House of Commons. This obstructive outcome would not only violate the stated purpose of the Senate as it currently exists, but could conceivably put the legislative body out of step with the value pattern of Canadian society.

For a government to be legitimate, the results of its output must be compatible with the values of the society it governs. At first glance it would appear that an elected Senate would be aligned with the values of Canadian society. After all, Canada is a democratic nation. This is what makes the concept of an elected Senate attractive. It's certainly not a hard sell to the public, especially when the appointed Senate has felt to many as unaccountable through numerous scandals over the years.

But is an elected Senate the best way to achieve legitimacy? The function of the Senate as a complimentary body to the House of Commons (O'Brien, 2019) must be taken into consideration; changes that threaten to undermine that role would certainly change the power dynamics of parliament in some unexpected ways, and it may also fail to fix the problem it aims to solve in the first place. The trouble with electing the Senate is that if Senators feel they are accountable directly to voters, they feel that they have a mandate of their own. Creating a second elected assembly creates the very real possibility of two competing democratic bodies that will seek to undermine one another. It creates a scenario where the Senate could see itself as being in competition with the House of Commons instead of being a complimentary body. If the Senate is competing with the lower chamber, it is not difficult to imagine it creating dysfunction within the legislative branch as a whole. Such dysfunction would certainly run counter to the values of Canadian society as expressed in the Supreme Court Reference (Hulme, 2016). In considering whether a democratic Senate would be in line with the values of Canadian society, we must look at the situation holistically. Balancing the popular will of the people in the moment with the long established traditions of a society is essential to determining its values. Further research would offer more clarity to these questions.

Conclusion

While the idea of an elected Senate, though easy to sell to voters, is not necessarily a better structure. Legitimacy is not automatically the result of democracy; as noted by Stillman (1974), it is important that we do not allow our conclusions to be skewed by ideological affiliations. I fear that this is a trap Canadian leaders have fallen into with regards to recent Senate reform attempts.

While nearly all the reform proposals brought forth since 1980 have adopted the idea of an elected Senate, it would seem that those that did failed to deeply examine the consequences of that course. It would certainly fix the legitimacy problem evident in the appointment process of Senators, but this solution also creates larger concerns with the essential function and purpose of the Senate. A likely outcome is that it would be a lateral move to a different sort of illegitimacy. We must be wary of trading one devil for another, particularly when, as Prime Minister Trudeau has so successfully demonstrated, there are easier solutions that can be attempted.

Perhaps the greatest lesson to be learned from the Trudeau reforms after 40 years of failed attempts, is that the best reform is about the art of the attainable. Even if one still believes that a democratic Senate is the ideal solution, it is not likely to happen because the process of constitutional amendment is too fraught with hazards. An imperfect solution today is better than a perfect solution that only exists in theory.

References

Bakvis, H. (2000). Prime Minister and Cabinet in Canada: An Autocracy in Need of Reform? Journal of Canadian Studies/Revue d'etudes canadiennes, 35:4, (pp 60-79) University of Toronto Press.

Docherty, D. (2002). The Cnadian Senate: Chamber of Sober Reflection or Loony Cousin Best Not Talked About. Journal of Legislative Studies, 8:3, (pp 27-48) Routledge, Taylor & Francis Group.

Hulme, K. (2016). Alberta's Great Experiment in Senatorial Democracy. American Review of Canadian Studies, 46:1, (pp 33-54) Routledge, Taylor & Francis Group.

O'Brien, G. (2019). Discovering the Senate's Fundamental Nature: Moving beyond the Supreme Court's 2014 Opinion. Canadian Journal of Political Science, 52:3, (pp 539-555) Cambridge University Press.

Stillman, P. (1974). The Concept of Legitimacy. Polity, 7:1, (pp 32-56) University of Chicago Press

"That's their decision": Hamilton's Policy and Protocol Response to Homelessness

Brianna Bedran

The pandemic presents an awakening for citizens in Hamilton of the pressing matter of housing struggles in Hamilton Ontario. Since the start of the pandemic, there has been a consistent and growing number of unsanctioned outdoor encampment facilities across the municipality. Although this may be a first sight for some, housing problems are deeply ingrained in the operation of our economic system and in the ways in which society functions, and they have not emerged in just the past few decades (Bratt et al., 2006). In Canadian cities, private-sector rents and residential real estate prices have increased at rates considerably higher than wages and salaries over the past ten years. These increases have unfolded with other economic forces and policy decisions, such as gentrification, deinstitutionalization, and reductions in welfare payments that can exacerbate serious housing needs (Collins, 2010). This is especially true for Hamilton as housing in Hamilton has risen to severely unaffordable rates making it the third least affordable housing market in Canada according to Oxford Economics (Oxford, 2021). The contrast of Canada's housing demand compared to supply dynamics has only worsened since the pandemic arised, with one of the worst cases being exemplified in Hamilton. This significant housing increase combined with the struggles of finding employment during COVID-19 leaving citizens little to no resources. In response to the obstacles people without housing face, Hamilton task forces have been dismantling encampment sites through collaboration with bylaw officers and police.

The quality of one's housing may also be an outward sign, as well as part of a person's self-image. Living in substandard housing in a bad neighborhood provides possible limits to people's ability to secure an adequate education for their children, reduces the chances of finding a

43

decent job and deprives them of finding a decent job and access to decent public services and community facilities (Bratt et al., 2006).

There are various ways anybody can experience homelessness, and the definition is not limited to outdated examples of one living on the streets. According to Collins, there are three types of homelessness: primary (people without shelter), secondary (individuals in numerous forms of emergency or short term shelter, including those temporarily staying with friends or relatives. Lastly, there are tertiary, classified as people living in single rooms and in private boarding houses without their own bathroom or kitchen or security of tenure. This group experiences a high degree of churning between the street and various forms of shelter (Collins, 2010).

The ability to provide comprehensive and quantitative answers to concerns of the size and characteristics of the homeless population is crucial to keeping homelessness on the political agenda, and to informing policy. (Collins, 2010). As recently released by Statistics Canada, more than 235,000 people in Canada experience housing struggles at any given year, and 25,000 to 35,000 people may be experiencing homelessness on any given night. Also gathered from administrative health data based on visits to emergency departments From 2010 to 2017, over 39,000 individuals were identified as experiencing homelessness. After 2014, the number of patients without housing visiting the ED increased until the last year of observation in 2017. These are just the statistics gathered based on people without housing who have visited hospitals, making it difficult to accurately represent just how many people are experiencing homleesness' as not all people reach out for assistance (Strobel et al., 2021). Narrowing down the statistics to our topic of focus, Hamilton, the Canadian Observatory for Homelessness reported the number of people without housing in the city stood at 385 in 2018 (Homeless Hub, 2021).

In 2013, Hamilton released their Approach to Addressing Affordable Housing & Homelessness. In this action plan, they did address the severity of homelessness in Hamilton and their mission to provide affordable housing for low-income families and individuals without housing, however in effect the plan was not adequate for the numbers of people experiencing homelessness.

Hamilton has since invested more in social housing, but even so the number of housing placements is over 100 people less than the amount experiencing homelessness. The number of people experiencing homelessness in 2018 was 385, and in the same year the number of placements was only 254. Additionally, the estimated wait time for social housing is nearly 2.5 years. If one finally gets placed in social housing, the average cost of rent for a 1 person bedroom apartment is $845 (Homeless Hub, 2021). This is a lower amount compared to average rent prices in the city, yet the average social assistance amount in Hamilton is $733 (Homeless Hub, 2021), which is not even enough to cover one's rent. It is clear that Hamilton's resources to combat homelessness are not allocated based on the data and are not representative of the amount of people experiencing housing issues. The number of placements is less than the number of people experiencing homelessness, the average wait time for a placement is way too long, and the contrast of social assistance to social housing rent cost leaves little to no space for extra funds.

Stephen Gaetz, Professor & Director of the Canadian Observatory on Homelessness, outlines the three common ways in which we can respond to homelessness. The approaches involve; prevention, emergency responses, and accomodation. Stopping people from experiencing homelessness in the first place, is of course the preventative response. Providing support like shelter, food and day programs, while someone is without housing is the emergency response, and the provision of housing and ongoing support as a means of moving people out of homelessness is accommodation. Hamilton has lacked on all three of these responses. Their policy response has focussed more so on the emergency responses and further accommodation in response to those emergencies. The city has neglected the first stage of prevention, as exemplified by their unsuccessful action plan. Moreover, their accommodation and emergency response to the crisis has been subpar at best.

The dismantling of encampments is an unjust response to homlessness that falls within the city's responsibility. As clarified by UN lead rapporteurs on the rights to adequate housing Kaitlin Schwan and Leilani Farha, Hamilton's protocol response to encampments neglects the rights and dignity of these residents. Further outlined by the International Human Rights Laws, governments are not permitted to destroy peoples' homes, even if those homes are made of improvised materials and established without legal authority (Farha, 2020). Thus, not only is the city creating further difficulties during a time of struggles, bylaw officers and police are revoking the rights of individuals trying to find shelter.

In August 2021, a year into the pandemic and task forces dismantling encampments, Hamilton released a new policy response that stated individuals will receive a fourteen day warning before assessment and removal will take place. Similar to an eviction notice, people living in encampments will have two weeks to find a new place to live before task forces will begin destroying their homes. Volunteer organizations such as the Hamilton Encampment Support Network and supportive Hamiltonians protested this response, putting pressure on the city to retract this protocol exactly a month later. Their new updated response is "six step plan" which is really just a longer version of saying that they will dismantle the encampments, but this time they will assess and provide accommodation for individuals in collaboration with Hamilton's Housing Services' Housing Focused Street Outreach Team. While it's an improvement the city is addressing their responsibility to provide accommodation of housing after tearing down people's homes, dismantling sites still denies their rights. Additionally, the apathetic attitude from the municipal government if people without housing are reasonably unhappy with these services is alarming.

Fred Eisenberger, the mayor of Hamilton, made some shocking and insensitive remarks in addressing the housing crisis. Don Mitchell reports the mayor told Global News that the refusal of services instead of outdoor encampments is "their decision" (Mitchell, 2020) and that these individuals "decide they want to find another way of living" (Mitchell, 2020). It is individual's without housing own choice, suggests the mayor, to struggle with secure living. Meanwhile, this past January Hamilton doctors clarified that the outbreaks of COVID-19 at shelters in Hamilton raises concern for more people experiencing homelessness to sleep rough. Hence, it is not their choice to find another way of living, the outbreaks and further closures of shelters are leaving little to no options (Rankin, 2021).

Eisenberger's insensitive comments, the lack of investment in housing services, and the dismantling of camps across the city demonstrates Hamilton is actively neglecting homeless individuals. Hamilton uses the emergency response approach to save face and implements last ditch policy efforts to respond to dire housing needs, and when presented with the increased severity of housing during COVID-19, their initial response was to dismantle their homes. Not only is adequate resources and funding necessary to respond to housing insecure individuals' struggles, a factor that should go without saying is upholding their dignity and rights.

References

A Right to Housing : Foundation for a New Social Agenda, edited by Rachel Bratt, et al., Temple University Press, 2006. ProQuest Ebook Central, https://ebookcentral.proquest.com/lib/mcmu/detail.action?docID=298850.

Collins, D. (2010). Homelessness in Canada and New Zealand: A Comparative perspective on numbers and policy responses. Urban Geography, 31(7), 932–952. https://doi.org/10.2747/0272-3638.31.7.932

Farha, L., et al. (2020) A National Protocol for Homeless Encampments in Canada. UN Special Rapporteur. https://www.make-the-shift.org/wp-content/uploads/2020/04/A-National-Protocol-for-Homeless-Encampments-in-Canada.pdf

Gaetz, S. (2014). Solutions: Prevention. Homeless Hub. https://www.rondpointdelitinerance.ca/blog/solutions-prevention

Hamilton. (2020, December 8). City of Hamilton Encampment Response. City of Hamilton, Ontario, Canada. Retrieved September 13, 2021, from https://www.hamilton.ca/social-services/housing/city-hamilton-encampment-response.

Hamilton. Hamilton | The Homeless Hub. (n.d.). Retrieved September 13, 2021, from https://www.homelesshub.ca/community-profile/hamilton.

Mitchell, J. (2020) Spike in Hamilton homeless encampments a 'difficult challenge,' says city. Global News. https://globalnews.ca/news/7204646/hamilton-homeless-encampments/

Rankin, C. (2021). Shelter outbreaks flag 'dire situation' for homeless in Hamilton, doctors say. CBC News. https://www.cbc.ca/news/canada/hamilton/shelter-hamilton-outbreaks-1.5877242

Statistics Canada. https://www.doi.org/10.25318/82-003-x202100100002-eng

Strobel, S., et al. (2021). Characterizing people experiencing homelessness and trends in homelessness using population-level emergency department visit data in Ontario, Canada.

PANEL 5: AN ACADEMIC LOOK
AT MODERN CULTURE

Understanding the Art of Piano Tone Quality

Terrence Wu[1], Alyssa Wu[2], Zachary Schauer[3], Peter A. Johnson[4],
John C. Johnson[4], Austin A. Mardon[4]

[1] University of Western Ontario, London, ON, Canada
[2] McMaster University, Hamilton, ON, Canada
[3] MacEwan University, Edmonton, AB, Canada
[4] University of Alberta, Edmonton, AB, Canada

Abstract

The mechanism of the modern piano has greatly evolved since the Middle Ages when the harpsichord was introduced. To properly execute the correct technique and achieve the best possible sound, the performer should incorporate speed of attack, flexibility in the arm and wrist, alignment in the hand, and overall body posture and weight distribution. The sound of the piano is heard from the vibration bouncing off the soundboard and can be artificially sustained by the damper pedal. Pressing down a piano key quickly will generate a powerful sound, however, this is not to say that attack speed is only used in loud passages, as it also becomes significant when playing quickly. Flexibility and bone alignment focuses on playing the piano comfortably while enabling the pianist to spend the least amount of energy to execute a passage of music.

Keywords: classical music, piano tone quality, performance, piano mechanisms, keyboard technique

Introduction

The modern piano was derived from the hammered dulcimer, which became the inspiration for the subsequent development of the clavichord, harpsichord, and pianoforte. The dulcimer was likely created around 900 AD in the Middle East (The Hammered Dulcimer, n.d.). The term is derived from the Latin and Greek words: dulce and melos, which when combined means "sweet tune" (Rossing, 2010). This instrument is a long wooden box with strings that were either plucked or struck with handheld hammers. This concept evolved into keys that could be pushed down, which would raise an object (i.e., the hammer or metal stick) to strike the string. To play the dulcimer either by plucking or striking the strings with a mallet, it was easier to play faster by pressing keys connected by a lever that would raise the metal tangent that would pluck the string.

The harpsichord and clavichord evolved based on this idea during the Middle Ages. The clavichord produces sound by striking iron or brass strings with small metal blades, which are called tangents (Rossing, 2010). The harpsichord employs a similar mechanism that was first developed around the 1500s and became unpopular during the 18th century. The core of the harpsichord's mechanism comprises little wooden jacks that vertically pluck the string. The disadvantage to the harpsichord is the lack of dynamic control. If the key is depressed with too much force, the jack will pop out of the frame (Truby, n.d.). Another disadvantage to the harpsichord and clavichord is that the strings would go out of tune quite easily because the strings are being pulled to create a pitch repeatedly; fortunately, these issues were all resolved with Bartolomeo Cristofori's invention of the modern piano in the 1700s. Cristofori named his invention after what the predecessors couldn't do -- soft (piano) or loud (forte) dynamics. His pianoforte uses hammers covered in felt (Stulov, n.d.), instead of metal tangents. On our modern piano, when the key is struck softly, the hammer will strike the string at a slower speed, which results in an overall softer sound. However, when the key is struck with more power, the hammer will hit the string faster, creating a louder sound (Conklin, 1996).

The four concepts that influence the sound being resonated from the piano include the attack speed, the wrist and arm flexibility, the alignment of the bones, and the posture and weight distribution of the performer (Methuen-Campbell, 1983). The attack speed refers to the speed at which the finger strikes the key, in terms of power and different types of music that use this technique. Flexibility focuses on playing the piano in the most efficient and comfortable way. Skeletal alignment is another essential to piano playing that ensures proper blood flow and minimal tension (Wheatley-Brown, n.d.). Posture is the link between flexibility and alignment and would make proper contact with the keys extremely difficult.

An Overview to Producing Tone on the Piano

Creating good piano sound depends on the way our hands and arms create fluid motions to play the piano in a relaxed manner and with the greatest degree of control. In particular, effective gestures are responsible for the control of the sound quality produced by the instrument (Massie-Laberge et al., 2019). Performers must be able to elicit the ideal tone qualities and then apply the correct techniques to deliver sound. The resulting string vibration bounces off the wooden soundboard, and transforms the vibration into audible sound waves, in turn, creating the pitches that listeners hear. Once a performer presses and holds a piano key, the sound will immediately begin to decay, because the vibration of the piano strings is slowing down. On top of the strings, for each key, there are dampers that will fall once the finger is lifted off the key and stop the sound. It is possible to sustain the sound by using the damper pedal or the sostenuto pedal. The damper pedal will keep the damper pads from stopping the vibration, and the player will not need to keep their finger pressed down on the key (Hirschkorn, n.d.). The sostenuto pedal will hold a singular damper pad and will only hold a singular key. A common misconception is that many people believe that a piano key produces the same sound, no matter how the performer strikes it. However, one of the most crucial technical elements is absorption with the wrist. There are multiple factors to striking the key that will change the sound (Methuen-Campbell, 1983).

Attack Speed

The attack speed of the finger determines how quickly the finger vertically strikes the key. This attack only uses the weight of the finger but is essential for faster tempos (Bernays & Traube, 2013). If the performer attempts to play a quick passage with maximum volume (i.e. loud) for long periods of time, their body cannot hold the high amount of energy required to execute proper technique.Bad technique may cause hand and wrist pain, poor blood circulation, and soreness in the head and/or neck (Wristen, 2000).

The modern piano, unlike its early predecessors, is the most sensitive to the speed of attack. The harpsichord would not be affected by how fast or slow the key is depressed as the plucking mechanism will not influence the speed at which the jack would pluck the string. This is because keys respond to how fast they are being pressed down and how fast the hammer strikes the strings (Truby, n.d.). This determines the tone quality of the sound produced and how soft or loud the sound is (Bella & Palmer, 2011). Playing piano with good tone quality, musical expression, and the ability to play at faster speeds with accuracy is the main goal for professional pianists. Good tone quality in performance helps to achieve a singing quality to the music, or in the words of Mozart, allow it to "flow like oil" (Chai, n.d.).

These fundamentals are achieved with how much the finger is raised before pressing the key. If the desired tone is powerful and loud, the finger would have to strike the key from a higher angle. On the contrary, if the performer wants to play a virtuosic passage, the fingers would have to stay closer to the keys. An analogy that can be used to describe this idea is regular-paced walking. When we walk, we generally take small steps at a time and we do not need to lift our feet high off the ground. Lifting our legs high up each time we take a step consumes a lot of energy, which isn't needed for us to walk. If we want to walk faster, we would take small steps and just lift our feet high enough to not trip on the ground. In this scenario, the action of our legs is similar to our fingers. When the pianist wants to play faster, they will keep their fingertips close to the keys to avoid moving long distances to play each key. This method saves energy, building endurance and strength.

Flexibility

By vertically striking the key with the speed of attack, we will hear a note, however, when the hammer hits the string so quickly with no control, the sound will be harsh. To maintain warm sound quality, the pianist needs to incorporate flexibility in the finger joints, the knuckles, the wrist and the arm. Generally speaking, there are two main categories of sound, one being bright, and the other being warm. For example, when playing a lullaby, the performer would usually strive for supple sounding tones, unlike a mazurka, which is characteristically a majestic Polish dance, which would usually require a bold and powerful sound. When the flexibility of the wrist and arm are used in combination with the speed of attack to produce a sound, the sound is much more expansive with considerably better quality sound. If we go back to our walking analogy, when we walk, we bend our ankles, our knees, and our hips. If we do not maintain flexibility, our range of mobility is limited. When we walk we shift our body weight from one leg to the other. The faster we want to run, the more flexible our joints must be. The same goes for piano technique, the performer must constantly shift their weight from one finger to the other but still bending the fingers and knuckles. Our fingers are somewhat easier to control since we can rotate our wrist to help us shift our energy.

Bone Alignment

Bone alignment in the hands, arms, and body is crucial for a strong and powerful sound. For the correct tone, the power must come from the player's arm or body weight (Laws, 2018). If the wrist becomes crooked or the arm is crooked, the attack of the key will only come from the player's finger, which is less powerful. Finger attack will be more suitable for playing softly and quickly, but loose wrists and good alignment are essential for playing loudly and striking multiple notes at a time. If the posture of the wrists and the hands are not loose and relaxed, the sound will become distorted because the muscles guiding the bones are crooked. For example, when forming a strong fist with your hand, notice how the wrist and hand are aligned. Then, make the same fist while bending the wrist downwards towards your elbow. The muscles are bent and lose their strength, and the tendons are unable to function properly. The same strong fist is impossible to achieve without proper bone alignment (Cordell, n.d.; Sakai, 2002).

Proper Posture and Weight Distribution in the Body

Piano posture begins with the height of the bench and "and adjusted so that [the] forearms are lined up with the key beds, that is, slightly below the surface of the keys themselves" (Mora et al., 2007). The performer is advised to sit on the edge of their chair, to assist with proper balance weight and easy accessibility to shifting body weight toward the keyboard when needed. Proper posture in combination with the alignment of the wrist and finger bones is the foundation of any good piano sound. The power must come from the lower back and sit bones, connected to the shoulder muscles, into the arm, wrist, hand and finger. The best posture focuses the least amount of energy when not needed. When the performer's joints and muscles are strained, the flow of weight and power cannot provide efficient gestures and movements.

Conclusion

Pianists should understand the fundamentals of piano mechanisms and the most effective way to produce the best sound. Our finger articulation and fluidity of movements and muscles. Playing the piano should be, for the most part, effortless, considering the anatomy of the hand, and the minimal amount of energy required to spend, using natural body and arm weight. In short, speed of attack, flexibility in the hand and arm, the alignment of the bones, and weight distribution are the basic fundamentals of proper piano performance technique and every pianist should incorporate these concepts into their everyday practice.

References

Bella, Simone Dalla, and Caroline Palmer. 2011. "Rate Effects on Timing, Key Velocity, and Finger Kinematics in Piano Performance." PLOS ONE 6 (6): e20518. https://doi.org/10.1371/journal.pone.0020518.

Chai, Congcong. n.d. "SELECTING FINGERING FOR PERFORMING MOZART'S PIANO MUSIC," 44.

Conklin, Harold A. 1996. "Design and Tone in the Mechanoacoustic Piano. Part I. Piano Hammers and Tonal Effects." The Journal of the Acoustical Society of America 99 (6): 3286–96. https://doi.org/10.1121/1.414947.

Cordell, Kristin N. n.d. "Piano Performance Injuries and Preventions," 20.

Laws, Catherine. 2018. "Moving Bodies, Piano Body." Performance Research 23 (4–5): 351–54. https://doi.org/10.1080/13528165.2018.1511034.

Massie-Laberge, Catherine, Isabelle Cossette, and Marcelo M. Wanderley. 2019. "Kinematic Analysis of Pianists' Expressive Performances of Romantic Excerpts: Applications for Enhanced Pedagogical Approaches." Frontiers in Psychology 9. https://doi.org/10.3389/fpsyg.2018.02725.

Methuen-Campbell, James. 1983. "Pianism." Edited by Heinrich Neuhaus. The Musical Times 124 (1689): 683–85. https://doi.org/10.2307/961428.

Mora, Javier, Won-Sook Lee, and Gilles Comeau. 2007. "3D Visual Feedback in Learning of Piano Posture." In Technologies for E-Learning and Digital Entertainment, edited by Kin-chuen Hui, Zhigeng Pan, Ronald Chi-kit Chung, Charlie C. L. Wang, Xiaogang Jin, Stefan Göbel, and Eric C.-L. Li, 4469:763–71. Lecture Notes in Computer Science. Berlin, Heidelberg: Springer Berlin Heidelberg. https://doi.org/10.1007/978-3-540-73011-8_73.

"Performance Practice Review | Journals at Claremont | Claremont Colleges." n.d. Accessed July 4, 2021. https://scholarship.claremont.edu/ppr/.

Rossing, Thomas D., ed. 2010. The Science of String Instruments. New York, NY: Springer New York. https://doi.org/10.1007/978-1-4419-7110-4.

Sakai, Naotaka. 2002. "Hand Pain Attributed to Overuse among

Professional Pianists: A Study of 200 Cases." Medical Problems of Performing Artists 17 (4): 178–80. https://doi.org/10.21091/ mppa.2002.4028.

Stulov, Anatoli. n.d. "Experimental and Theoretical Studies of Piano Hammer." In Proceedings of SMAC 03, 175–78.

"The Hammered Dulcimer." n.d. Smithsonian Institution. Accessed July 4, 2021. https://www.si.edu/spotlight/hammered-dulcimer.

Truby, Roy. n.d. "Elementary Harpsichord Technique," 3.

Wristen, Brenda G. 2000. "Avoiding Piano-Related Injury: A Proposed Theoretical Procedure for Biomechanical Analysis of Piano Technique." Medical Problems of Performing Artists 15 (2): 55–64. https://doi. org/10.21091/mppa.2000.2012.

DAY 2

PANEL 1: DEVELOPMENTS IN BIOLOGY

Screening and Management Outcomes for Tuberculosis in Northern Communities

John Christy Johnson

Tuberculosis (TB) is a stealthy and prevalent disease in the North, using latent forms to be transmitted between individuals, communities, and even wildlife. This article seeks to evaluate the strengths and weaknesses of current screening and management protocols for TB in Arctic communities.

Several studies have looked at screening/diagnosis of latent TB as an area of both success and improvement (Pease et al., 2019). The tuberculosis skin test (TST) represents the recommended modality for screening TB and is being performed post-exposure to active TB cases or following suspicious symptoms. While the screen itself is simple, cost-effective, and quick at revealing a diagnosis with relatively high sensitivity and specificity, there are still diagnoses that slip under the radar. For instance, providing sputum induction for TB diagnosis in high-incidence Arctic communities was associated with cost-saving and community-based approaches were associated with greater effectiveness (Sugarman et al., 2014).

More rigorous testing and screening exams are relatively scarce in places like Nunavut and require transportation to a better equipped facility. Advanced medical interventions are a resource-exhaustive process as the primary and secondary care associated with a TB infection includes the time, money, and energy required for medical transport to facilities in the South (Alvarez et al., 2014). Additionally, health care interventions including diagnostic testing, nursing and physician assessment and TB medication are generally covered, although nationalistic and bureaucracies may be involved.

Professionals have reported that TB therapeutic management and symptom management outcomes ought to be more carefully examined. With the incidence of TB being the highest among the Inuit in Canada, treatment with rifapentine and isoniazid once weekly for 12 weeks has been made accessible for certain latent TB infections (Alvarez et al., 2020). However, medication adherence is one of the major factors that affect outcomes and prognosis and there is evidence to suggest that longer and more complex medication regimens can increase likelihoods of non-adherence (Pradipta et al., 2020). As such, under resourced community health facilities have greater responsibilities to educate, counsel, and develop service for these patients.

Lastly, there has been a suggested role for the scrutinization of transmission control or prevention strategies. Evidence was largely sparse but appear to suggest a role for immunizations, the stratification of latent TB infections using interferon-γ release assay screening, medication adherence, and the de novo emergence of multi-drug-resistant TB (Alvarez et al., 2014).

There is currently limited evidence that indicates strengths of remote Arctic TB mitigation include immunization, risk stratification, and community initiative, while there remain limitations for these areas like cost, program feasibility, implementation logistics, and recurrent disease.

References

Pease, C., Zwerling, A., Mallick, R. et al. (2019). The latent tuberculosis infection cascade of care in Iqaluit, Nunavut, 2012–2016. BMC Infect Dis 19, 890. https://doi.org/10.1186/s12879-019-4557-3

Sugarman J., Alvarez G.G., Schwartzman K. et al. (2014). Sputum induction for tuberculosis diagnosis in an Arctic setting: a cost comparison. Int J Tuberc Lung Dis, 18(10):1223-30. doi: 10.5588/ijtld.14.0163. PMID: 25216837.

Alvarez, G. G., Van Dyk, D. D., Davies, N., et al. (2014). The feasibility of the interferon gamma release assay and predictors of discordance with the tuberculin skin test for the diagnosis of latent tuberculosis infection in a remote Aboriginal community. PloS one, 9(11), e111986. https://doi.org/10.1371/journal.pone.0111986

Alvarez G. G, Van Dyk D., Mallick R., et al. (2020). The implementation of rifapentine and isoniazid (3HP) in two remote Arctic communities with a predominantly Inuit population, the Taima TB 3HP study. Int J Circumpolar Health. 79(1):1758501. doi: 10.1080/22423982.2020.1758501. PMID: 32379538; PMCID: PMC7241515.

Pradipta I.S., Houtsma D., van Boven J.F.M., et al (2020). Interventions to improve medication adherence in tuberculosis patients: a systematic review of randomized controlled studies. NPJ Prim Care Respir Med. 30(1):21. doi: 10.1038/s41533-020-0179-x. PMID: 32393736; PMCID: PMC7214451.

The Best Online Teaching and Learning Practices to Implement While at Home

Alyssa Wu[1], Peter Johnson[2], John Johnson[2],
Jasrita Singh[3], Zach Schauer[4], Austin Mardon[2]

[1] University of Western Ontario, London, ON, Canada
[2] University of Alberta, Edmonton, AB, Canada
[3] McMaster University, Hamilton, ON, Canada
[4] MacEwan University, Edmonton, AB, Canada

Abstract

Billions of people around the world are facing the challenges brought on by the COVID-19 pandemic. This all started when a new coronavirus strain (SARS-CoV-2 virus) was found in patients with pneumonia-like symptoms, located in Wuhan, China. During this time of quarantine and self-isolation at home, many students are continuing to learn and motivate themselves to finish their studies online. Students and teachers from all around the world have had to accommodate their learning and teaching styles to be suitable for learning at home. Modern education practices now heavily rely on having adequate internet access, for sending and receiving information. Online communication platforms are being used to facilitate student-teacher interactions. Teachers are also finding better ways to share and communicate course materials effectively over the internet, in order for their students to benefit and supplement their learning progress. Educators around the world have been adapting their in-person learning environments to a similar online environment as smoothly as possible, given the current circumstances of the global COVID-19 pandemic, surrounding lockdown procedures, quarantine measures, and travel restrictions. This review will focus on the challenges in various learning sectors, ranging from elementary, secondary, and post-secondary education. There will also be a focus on improving teaching practices to benefit students and help them learn and engage in more material while at home.

The Current State of Online Learning

Online platforms are now being used to recreate the human-to-human interactions that most students are used to having in their day-to-day learning before the lockdown measures were enforced by the COVID-19 pandemic. As internet usage increases, supplementing online learning with in-person learning, known as blended learning, is becoming more prevalent (Singh, 2003). Given the current state of the world, instructors are now having to alter their teaching styles and use more online learning. While there are advantages and disadvantages to online learning, there is still a lot that can be done to supplement and improve students' learning experiences.

Currently, many forms of online learning require students to be motivated, self-directed, and goal-oriented. Students will use more materials (such as readings, videos, etc.), rather than more personal interactions (such as discussions, presentations, etc.). Communication is vital in online learning (Rapanta et al., 2020). Teachers should try to keep students on track with a schedule laying out all the materials that students should be going through weekly. This also serves to keep students motivated and engaged with the material. Many instructors have also found it to be effective for students when reminders are sent out prior to assignment deadlines, term tests, etc. In addition, announcements that describe when new content/modules have been released are helpful for students to keep track of their progress throughout the course.

However, with the decrease in direct communication from the instructor's end, many students find it difficult to absorb the material that they are learning. In a normal in-person learning environment, students are used to engaging in teacher-centred learning, where what they learn is directly influenced by the resources that the teacher provides (Emaliana, 2017). When students go through successive modules filled with new material, it can be very overwhelming for the student as some may find it difficult to grasp main concepts or ideas being taught that are not directly communicated from the instructor themselves. This factor can lead to students losing focus in their coursework, and they may miss the objectives of the new material being taught.

The Use of Online Communication Platforms

Online communication platforms have long existed, but their use has heavily increased during the pandemic. Individuals from all around the world are finding new and creative ways to facilitate the sessions that they would normally hold in-person. Of course, with a significant increase in usage of these everyday communication tools, comes many more ideas that facilitate interesting and perhaps more engaging ways of hosting or participating in an event. There is a wide range of online audio-video communication platforms that are now being used (eg. Zoom, Skype, Google Classroom, Microsoft Teams, Cisco Webex etc.) to facilitate and simulate the transition of being in a classroom to being at home, with a computer in hand (Sahu, 2020).

Many online communication platforms have designed new features to simplify the online teaching and learning experience. Educators can take advantage of screen-sharing and annotation functionalities, where they can prepare a presentation and some accompanying videos as a visual to help students learn while they are teaching and explaining a new idea or novel concept. Instead of a chalkboard/whiteboard, teachers now have the option to use digital whiteboards to write and solve problems with their students' on a live call. Some platforms (such as Zoom) enable students to annotate the screen as well, and it might be engaging for everyone in the class if the instructor would like to have some students volunteer and share their ideas. Of course, a downside to utilizing some of these great functionalities is that they are limited depending on the devices that are accessible for students and teachers.

Elementary and Secondary School Education

Different school boards have been designing and using their own platforms to enable easy access for parents and students to obtain learning resources to support their learning (Mulenga & Marbán, 2020). Secondary schools and high schools have tried various methods to help transition their students into this new learning environment while supporting them and preparing them for their future endeavours - whether their goal is to pursue higher education, join the workforce, etc. (Wang, 2013). This period of change has not been entirely smooth along the way, and some students have reported feeling 'unprepared' in pursuing their future career aspirations.

To support the transition from in-person school to online school, there has been an increase in making online digital libraries, which contain books, worksheets and additional learning materials for parents to download for their children. However, the effectiveness of these online libraries depends on an increase of awareness to better integrate these online resources into student learning (Sharifabadi, 2006). However, despite the continuous efforts in making this transition as smooth as possible, there are definitely some limitations in being able to keep young children (especially those at the elementary level) engaged in completing their school work entirely online. Many children do not have access to their own personal devices, and so, they will need to rely on parental help and the use of their devices to complete their assigned schoolwork. But as more and more parents return to work, this is not always a feasible option. Younger children need the guidance and structure provided by parents and teachers, in order to succeed. With increasing class sizes directly relating to limiting time constraints, it is very difficult for teachers to provide one-on-one support for children who need it most (Mulholland & O'Connor, 2016).

Education in Universities/Colleges

Many professors have been hard at work, transitioning their lectures to be in a user-friendly online format for their students. With this factor in mind, there has definitely been a wide range of challenges as it is difficult to fully simulate a lecture-style learning environment through a digital screen.

Numerous institutions are starting to default to pre-recorded lectures. This is often easier for instructors in many circumstances, as it eliminates the potential glitches in technology that can occur during a live lecture with hundreds of students tuning in on a live call (e.g., facilitated through Zoom). However, from a student's perspective, there are definitely some benefits and downsides to the use of asynchronous, pre-recorded lectures. Pre-recorded lectures allow students more flexibility in watching these lectures on their own time. These types of lecture styles require a great deal of time management on the student's behalf. Taking an opposing standpoint, students tend to fall behind in these lectures, as they are often unscheduled, and up to the student's discretion on when to watch them. The use of small assessments (such as quizzes and assignments) should be added to the end of each week, as an additional incentive to keep students on track and engaged while completing their course work (Kamal et al., 2020).

Synchronous (or live) lectures can sometimes be more engaging for professors and students alike, but it also adds another layer of stress. Technological challenges (such as internet stability and bandwidth) can cause delays and lags, which disrupt the flow of information being transmitted from professors to students (Gillett-Swan, 2017).

General Teaching Recommendations and Suggestions

Designing a successful online learning environment should focus on a student-centred design. Instructors should keep in mind that the content that they are putting up for students to view should be of good quality, and they should benefit the students in learning the desired material (Rapanta et al., 2020). There are many options that are readily available to help instructors develop good quality learning resources, such as podcasts, pre-recorded lectures, educational videos, article/textbook readings, etc. In addition, instructors should keep in mind the practicality of the resources that they are sharing with the class. Time is valuable for everyone, and many students have a wide range of other commitments that they have to focus their attention on -- including schoolwork, personal life, finances, etc. Instructors should try to filter through their resources, and assign only the necessary sections that contain the most valuable information relevant to the course material, for the benefit of the students. For example, when looking at a long article, the instructor could skim through and guide their students towards the most important sections - rather than having them read the entire article. This is a more efficient method, as it saves the student some time, and also prevents distractions from having the student read a lot of extraneous material (Förster et al., 2018).

When teachers are facing these problems or a lack of engagement from their students, they can consider switching up their teaching styles to better accommodate the student's immediate struggles. Be careful not to be too pushy or persistent in achieving the goals of a given task. Sometimes, it might be better to completely switch over to another piece of material, or try a different activity altogether. When instructors are constantly changing up the structure of their lessons, this can keep the students engaged for a longer period of time.

Teachers should also consider creating interactive presentations and puzzles for the students to solve in class. One of the most important ideas to keep in mind while running an online class is the level of student engagement. Some teachers have opted in and out of having students turn on their cameras, but with that, this presents another set of challenges to overcome. At home, it is harder to control the number of distractions, as every household is unique. When compared to an in-person classroom, general distractors can be easily eliminated by the teacher.

For teachers running classes with older students (grades 4+), dedicating some time to open discussion may help improve student engagement. Teachers can call upon a student who has their 'hand raised' to share their thoughts and opinions on a given question. Another option is to utilize the "chat" feature that is embedded in many digital platforms. The teacher could recruit an assistant, who will be able to monitor the incoming chat messages during class. The assistant could serve to filter and report back on frequently asked questions and summarize general comments and remarks made by the students in the class.

References

Andersen, Kristian G. et al. "The proximal origin of SARS-CoV-2". Nature Medicine, vol. 26, Mar. 2020, pp. 450-452. https://doi.org/10.1038/s41591-020-0820-9.

Emaliana, Ive. "Teacher-centered or Student-centered Learning Approach to Promote Learning?". Jurnal Sosial Humaniora, vol. 10, no. 2, Nov. 2017, pp. 59-70. http://dx.doi.org/10.12962/j24433527.v10i2.2161.

Förster, Natalie, et al. "Short- and long-term effects of assessment-based differentiated reading instruction in general education on reading fluency and reading comprehension". Learning and Instruction, vol. 56, Aug. 2018, pp. 98-109. https://doi.org/10.1016/j.learninstruc.2018.04.009.

Gillett-Swan, Jenna. "The challenges of online learning: supporting and engaging the isolated learner". Journal of Learning Design, vol. 10, no. 1, 2017, pp. 20-30. https://eprints.qut.edu.au/102750/.

Kamal, Ahmad A. et al. "Transitioning to Online Learning during COVID-19 Pandemic: Case Study of a Pre-University Centre in Malaysia". International Journal of Advanced Computer Science and Applications, vol. 11, no. 6, 2020, pp. 217-223. https://philpapers.org/rec/KAMTTO-8.

Mulenga, Eddie M. and Marbán, José M. "Prospective Teachers' Online Learning Mathematics Activities in The Age of COVID-19: A Cluster Analysis Approach". EURASIA Journal of Mathematics, Science and Technology Education, vol. 16, no. 2, May 2020, pp. 1-9. https://doi.org/10.29333/ejmste/8345.

Mulholland, Monica and O'Connor, Una. "Collaborative classroom practice for inclusion: perspectives of classroom teachers and learning support/resource teachers". International Journal of Inclusive Education, vol. 20, no. 10, Feb. 2016, pp. 1070-1083. https://doi.org/10.1080/13603116.2016.1145266.

Rapanta, Chrysi, et al. "Online University Teaching During and After the Covid-19 Crisis: Refocusing Teacher Presence and Learning Activity". Postdigital Science and Education, July 2020. https://doi.org/10.1007/s42438-020-00155-y.

Sahu, Pradeep. "Closure of Universities Due to Coronavirus Disease 2019 (COVID-19): Impact on Education and Mental Health of Students and Academic Staff". Cureus, vol. 12, no. 4, Apr. 2020. https://doi.org/10.7759/cureus.7541.

Sharifabadi, Saeed R. "How digital libraries can support e-learning". The Electronic Library, vol. 24, no. 3, May 2006, pp. 389-401. https://doi.org/10.1108/02640470610671231.

Singh, Harvey. "Building Effective Blended Learning Programs". Educational Technology, vol. 43, no. 6, Dec. 2003, pp. 51-54. https://www.ammanu.edu.jo/EN/Content/HEC/6.pdf.

Wang, Xueli. "Why Students Choose STEM Majors: Motivation, High School Learning, and Postsecondary Context of Support". American Educational Research Journal, vol. 50, no. 5, Oct. 2013, pp. 1081-1121. https://doi.org/10.3102%2F0002831213488622.

Inhibition of MALT1 by Mepazine Acetate for the Treatment of Psoriasis Induced by CARD14 Mutations

Tara Y.T. Chen[1] and Meera Chopra[2]

[1] Schulich School of Medicine and Dentistry, Western University
[2] Faculty of Health Science, McMaster University

Abstract

Psoriasis is a chronic inflammatory skin disease characterized by red and scaly patches on the epidermis. It results from overactive immune responses induced by the NF-κB protein complex, which increases proinflammatory gene expression. As a result of increased inflammatory responses, keratinocytes in the epidermis are overproduced and accumulate, creating psoriatic plaques. One cause of this overactive immune response is attributed to genetic mutations in the caspase recruitment domain-contain protein 14 (CARD14), which causes overactivity of NF-κB via amplified enzymatic activity of the mucosa-associated lymphoid tissue lymphoma translocation protein 1 (MALT1). Topical therapies for psoriasis do not significantly alleviate symptoms for severe cases, however, the systemic inhibition of MALT1 may be more effective. One potential therapeutic compound is mepazine acetate, a phenothiazine derivative that allosterically inhibits MALT1.

The impact of CARD14 mutations on the development of psoriasis was modelled using the COBWEB simulation software. COBWEB was then used to model the inhibition of MALT1 by mepazine acetate by comparing the control with treatment models. The simulations with the mepazine acetate treatment had a reduced accumulation of dead keratinocytes and decreased proliferation of psoriatic plaques in the upper epidermal layers, compared with control simulations after seven days. This model is an idealistic representation of psoriasis; variation between these results and clinical findings is likely. These results demonstrate that mepazine acetate decreases MALT1 activity, thus reducing proinflammatory responses.

The next steps include building a more complex model with the ability to substantially vary responses to treatment.

Keywords: psoriasis, mucosa-associated lymphoid tissue lymphoma translocation protein 1 (MALT1), caspase recruitment domain-contain protein 14 (CARD14)

Introduction

Psoriasis is a chronic inflammatory skin disease that appears as red and scaly patches on the epidermis. It is caused by unsuitable immune responses in the body, resulting in inflammation, dilated capillaries, and plaque buildup (Alshobaili et al., 2010). Psoriasis can appear at any age, however, genetic and environmental factors affect individual susceptibility and presentation (Kamiya et al., 2019). Certain alleles located in the Major Histocompatibility Complex cause a 60-90% chance of heritability of the disease (Alshobaili et al., 2010). As well, environmental triggers may cause psoriasis to appear in genetically predisposed individuals. For instance, extrinsic environmental triggers include mechanical stress induced by physical injuries, UV exposure, beta-blocker drugs, and air pollution. Intrinsic environmental factors include obesity, smoking, alcohol, hypertension, metabolic syndromes, and stress (Kamiya et al., 2019).

Mutations to caspase recruitment domain-contain protein 14 (CARD14) are associated with the pathogenesis of psoriasis. This intracellular gene is located in the PSORS2 locus, which acts as a scaffolding protein and is involved with proinflammatory gene expression. It forms a signalling complex with B-cell lymphoma/leukemia 10 (BCL10) and mucosa-associated lymphoid tissue lymphoma translocation protein 1 (MALT1) that activates NF-κB signaling. This CARD14-BCL10-MALT1 (CBM) complex is enhanced in the presence of CARD14 mutations, such as p.Glu138Ala, p.Glu142Lys, p.Gly117Ser, and p.Glu142Gly, resulting in an overactivation of NF-κB and excess production of chemokines such as CXCL8 and CCL20 (Van Nuffel et al., 2017). As a result, keratinocytes are overproduced, creating psoriatic plaques (Van Nuffel et al., 2017). In addition, other paracrine mediators such as interferons and Th17 cytokines can activate the Dectin-1 G-protein coupled receptor (GPCR) on the cell membrane of keratinocytes, resulting in the production of more CBM complexes, overactivation of NF-κB, and increased MALT1 protease activity (Juilland & Thome, 2018).

Since the proteolytic activity of MALT1 in keratinocytes due to CARD14 variants in CBM complexes contributes to the expression of psoriasis, the inhibition of MALT1 may help alleviate disease symptoms. A noncompetitive MALT1 inhibitor is mepazine acetate, a phenothiazine derivative that interacts with the allosteric pocket of MALT1. In mouse models with multiple sclerosis, decreased inflammation in the central nervous system was observed after treatment with mepazine acetate (Mc Guire et al., 2014). This compound was also found to suppress peripheral autoreactive T cell activation in vitro, inhibiting the production of IL-2, IFNγ, and IL-17 cytokines, as well as hindering TCR-induced signalling involved with immune responses related to the disease. Peripheral regulatory T cells, which prevent autoimmune disease and regulate immune responses, were not affected by mepazine acetate treatment (Mc Guire et al., 2014). Additionally, when mepazine acetate was used as an antipsychotic drug under the brand name Pacatal in the 1950s, no long-term side effects were revealed after using the treatment at concentrations between 100 and 800 mg four times a day (Mc Guire et al., 2014). Research suggests that administration of mepazine acetate to cells at a concentration of 4.5 μg/mL (13 μM) every two days can effectively block unfavourable MALT1 protease activity in T cells (Meloni et al., 2018).

Topical corticosteroids are commonly used for mild to moderate cases of psoriasis. They upregulate the transcription of genes with anti-inflammatory functions, including the IL-1 receptor antagonist, IL-10, and tyrosine aminotransferase (Uva et al., 2012). However, patients who receive topical therapy often report minimal symptom improvement (Martin et al., 2019). Another therapy for more proliferative and severe cases of psoriasis is ultraviolet (UV) phototherapy. Research suggests that UVB light causes immunosuppression due to the induction of apoptosis in both keratinocytes and T cells, which are involved in pathways causing psoriasis (Zhang & Wu, 2018). Current therapies that are being developed include the use of biologic agents for the treatment of psoriasis, which, for example, compete for the binding site of TNF-α on TNF receptors 1 and 2. The actions of TNF-α typically activate the proliferation of macrophages and stimulate the release of interferon-β, resulting in the characteristic inflammation of psoriatic plaques (Schadler et al., 2019).

In addition, systemic MALT1 inhibition with mepazine acetate may improve the prognosis of severe psoriasis cases more effectively compared to topical therapies (Van Nuffel et al., 2017). This compound can be administered systemically through the use of transdermal patches, which provide a specific dose of medicine through the skin (Michigan Medicine). The drug travels through the epidermis, dermis and hypodermis without accumulation in the upper skin layers and reaches the stratum basale of the epidermis where the drug interacts with MALT1 (Mayo Clinic, 2021). For a treatment to be delivered through a transdermal patch, it must be in a gel form and lipid-soluble with a molecular weight below 500 Da (Pastore et al., 2015). Depending on the initial amount and concentration of the treatment in the patch, the treatment will be released into the skin at a constant rate. However, the concentration of the treatment inside the patch decreases the longer it is worn, which results in the drug being delivered at a slower rate. Transdermal patches are usually replaced every few days to ensure that the treatment is delivered at the appropriate rate (Pastore et al., 2015). The chemical properties of mepazine acetate are presented in Table 1.

Using a computer simulation, the therapeutic effect of a transdermal patch containing mepazine acetate on the accumulation of keratinocytes in the epidermis was measured. Given the proven effect of mepazine acetate to allosterically inhibit MALT1, a reduction in keratinocyte accumulation in the stratum corneum is hypothesized to occur.

Methods

COBWEB

COBWEB is a simulation-based software modelling social, economic, and biological systems. Agents, the primary decision-makers in COBWEB models, are represented by triangles with a coloured dot in the centre to differentiate between agent types. Their primary sources of energy are solid-coloured squares, named resources. Based on different parameters set by researchers to create unique environments, agents make different decisions regarding moving, collecting resources, and breeding. Ticks are an arbitrary unit of time. The COBWEB model is shown in Figure 1.

Parameters

Agent Abiotic Factors

The Agent Abiotic Factors tab was used to split the 100 by 100 grid into three layers. The bottom section represents the stratum basale of the epidermis, the middle represents the stratum spinosum, granulosum, and lucidum layers, and the top represents the stratum corneum.

Agents

70

Four agents are in use in this model, each with a coloured dot to differentiate between them. The blue agents, representing keratinocytes, produce CBM complexes after the activation of the Dectin-1 GPCR by interferons and Th17 cytokines (de Koning et al., 2010). In the model, cells that are activated to produce CBM complexes are "diseased" and present with a blue colouring. Diseased agents are able to pass on the disease via contact with other agents. These agents are confined to the lower layer of the model.

The yellow agents represent polygonal keratinocytes. They can be induced with the "disease" through contact with diseased blue agents to produce CBM complexes. These agents are confined to the middle layer of the model.

The green agents represent dead keratinocytes present in the stratum corneum. After contact with diseased yellow agents, these agents age at a quarter of the rate of their healthy counterparts, resulting in a buildup of dead cells on top of the skin. This buildup characterizes psoriatic plaques. These agents are confined to the upper layer of the model.

The red agents represent the mepazine acetate treatment and are placed in the lower layer of the model. These agents represent an allosteric inhibitor of the MALT1 present in CBM complexes (Mc Guire et al., 2014). These agents are initially placed in the lower layer of the model, but are permitted to move freely within the grid.

Blue, yellow, and green agents are set with an age limit to represent the life span of keratinocytes. Agents are initially present in small quantities and allowed to proliferate until 2,500 ticks. At this point, the model of the skin is fully developed and plaques are visible at the top layer.

Swarming

To closely mimic psoriatic plaques on the stratum corneum, the "swarming" function of dead keratinocytes was enabled. Agents group together, demonstrating inflamed areas that are characteristic of psoriasis.

Disease

In the disease tab, agents are allowed to transfer the "disease" of producing CBM complexes to one another. Red agents are modelled as "healers" and after they come into contact with diseased agents, the latter are "healed" and do not propagate the "disease" further.

AI

The AI tab controls predetermined settings of how agents and the environment will interact with one another. Between trials, the AI seed was randomized to mimic the diverse presentations of psoriasis.

Control Simulation

Agents are allowed to proliferate freely for 37,500 ticks (data analysis starts at 2,500 ticks to allow for cell proliferation and plaque formation). In this model, there are no treatment (red) agents.

Treatment Simulation

Treatment simulations use the same parameters as the control simulation. However, at 2,500 ticks, red agents are inserted into the lower layer of the model to represent treatment by means of a transdermal patch. The model is allowed to run for 35,000 more ticks, representing a full week of treatment. Twenty-five control and treatment trials were conducted, using the same AI seeds for each trial pair.

Transdermal Patches

Transdermal patches were chosen as the treatment diffusion method of this model. Due to the micro-needles on transdermal patches, treatment agents were placed in the lower levels of the epidermis and allowed to move around the grid. Other pharmaceutical products that are administered via transdermal patches, such as clonidine and oestradiol, have patches that are replaced every seven days (Michigan Medicine). As a result, the psoriasis model runs for the first seven days of treatment and is intended to have a replenishment of treatment agents after 35,000 ticks.

Statistical Analysis

A two-sample t-test assuming equal variances was conducted between the average amount of CBM complex-producing dead keratinocytes at 37,500 ticks of control and treatment trials. A p-value obtained and compared to an alpha value of 0.05.

Results

The purpose of this study is to determine the effect of a simulated treatment on the proliferation of psoriatic plaques. The amount of CBM complex-producing dead keratinocytes, representing cells that would contribute to plaque buildup, in the treatment trial decreased over time with a slope of -0.0018 (Figure 2.). The amount of CBM complex-producing dead keratinocytes in the control trial increased over time with a slope of 0.0002. The mean total number of CBM complex-producing dead keratinocytes present in the stratum corneum after 35,000 ticks was measured, with the average result of control trials on the left (blue) and the average result of treatment trials on the right (red) (Figure 3.). On average, control trials had an average of 662.444 CBM complex-producing dead keratinocytes contributing to plaque buildup, whereas treatment trials had an average of 619.852 contributing cells. The number of CBM complex-producing dead keratinocytes in treatment trials was significantly lower than in control trials ($p < 0.05$) (Figure 3).

Discussion

Psoriasis is a skin condition characterized by the buildup of keratinocytes in the upper layers of the epidermis, resulting in red and scaly patches on the upper surface. Current treatments are applied topically and can have varying levels of success depending on the patient. The main factor preventing the effectiveness of topical therapies for the treatment of psoriasis is the epidermal permeability barrier. Compounds above a molecular weight of 500 Da are not able to enter epidermal layers, and given that some corticosteroid treatments weigh above 800 Da, they would not be able to enter the skin to induce immunosuppressive effects (van de Kerkhof et al., 2011). Transdermal patches circumvent this issue, as they present the treatment at the lowest epidermal levels through microneedles. As a result, treatment can be delivered effectively to cells (Pastore et al., 2015).

Transdermal patches may be placed on the upper body and upper arms on dry, hairless skin. The treatment is present in the patch in a gel form and enters systemic circulation at a constant rate. The therapeutic drug travels through the skin without accumulation in epidermal layers and disperses throughout the body via the circulatory system (Alkilani et al., 2015). The rate of replacement of the patches depends on the initial concentration and the body's rate of metabolism of the treatment (Martin et al., 2019). Transdermal patches are useful for psoriasis patients who have moderate to severe levels of psoriasis and have not had success with topical treatments (Martin et al., 2019).

Transdermal patches are used for transdermal drug delivery for a variety of purposes such as female hormone therapy, treating pain with fentanyl, alleviating symptoms of Parkinson's disease and nitroglycerin delivery to prevent angina for patients with coronary artery disease (Pastore et al., 2015).

There is currently a patent for the usage of the (S)-enantiomer of mepazine to inhibit a paracaspase, specifically MALT1 (Krappmann et al., 2017). It can be used to target diseases that can be treated by MALT1 inhibition. For instance, the (S)-enantiomer of mepazine can treat cancers (eg. lymphoma) and autoimmune diseases (eg. multiple sclerosis). Moreover, in a 2014 study, mepazine acetate was used to treat a murine model with multiple sclerosis (Mc Guire et al., 2014). This treatment reduced multiple sclerosis symptoms and peripheral autoantigen-specific T cell responses. However, it did not affect the development of peripheral regulatory T cells, which are important for immune system homeostasis and preventing autoimmunity (Mc Guire et al., 2014).

In this model of moderate to severe psoriasis, mepazine acetate was delivered using a transdermal patch with the intent of patch replacement every seven days. To simulate the transdermal patch treatment of psoriatic plaques, the COBWEB software was used to model the various layers of the skin and cell-to-cell interactions. Three agents were used in the model to represent various types of keratinocytes and a final agent was used to represent an allosteric inhibitor of the MALT1 present in CBM complexes, mepazine acetate. The initial three agents that are endogenous to the skin are allowed to proliferate for 2,500 ticks, an arbitrary measure of time used in the COBWEB software. Treatment and control models are run for a subsequent 35,000 ticks, representing one week.

Trials with the mepazine acetate patch treatment had fewer CBM-producing keratinocytes and a reduced accumulation of keratinocytes in upper epidermal layers compared to control trials after seven days. These results demonstrate that preventing the overactivation of the NF-κB signalling pathway and thus decreasing MALT1 proteolytic activity, effectively prevents the formation of psoriatic plaques. This systemic treatment mechanism may be more effective than topical corticosteroids, as it enters the circulation and proliferates throughout the entire body. Although there was a moderate decrease in the accumulation of keratinocytes after seven days, long-term usage of the patch may create a larger reduction of proliferation.

In a 2017 study, a Matlab model was built to assess the therapeutic impact of blue light irradiation on psoriasis (Félix Garza et al., 2017). Both healthy and diseased keratinocytes interacted with one another in a bi-stable system, and the hyperproliferative nature of psoriatic skin was modelled using 12 mathematical equations representing the kinetics of keratinocyte interactions. Similarly to the COBWEB model, the Matlab model represented interactions across epidermal strata between diseased and non-diseased keratinocytes. Psoriatic plaques were created in both models using rapidly proliferating cells. The study observed no therapeutic differences in psoriasis severity when the simulated skin was treated with low versus high intensity blue light, suggesting that duration of blue light exposure determines treatment efficacy (Félix Garza et al., 2017). Although a dose-response relationship was not measured in the COBWEB model, determining this relationship is a promising area of future research.

Conclusion

The COBWEB model determined that mepazine acetate may be administered systemically through a transdermal patch to treat moderate to severe levels of psoriasis. Common treatments for psoriasis that are applied topically have limited preventative effects and may not be feasible for patients with widely spread and severe psoriatic plaques (Uva et al., 2012).

A limitation of this study was that the COBWEB model is an idealistic representation of psoriasis presentation in patients. Since clinical cases rarely, if ever, present as ideally as software models, there is bound to be some variation between the results shown in this model and clinical findings. As a result, findings may not be representative of all clinical cases. Future directions of this study include using more complex modelling software to represent individually stratified layers of the skin as well as more agents that could represent immune mediators of the psoriasis pathway.

Acknowledgements

This research was conducted with the help of Dr. Brad Bass and members of the University Research Experience with Complex Systems at the University of Toronto. We thank them for their support and guidance.

References

Alkilani, A. Z., McCrudden, M. T., & Donnelly, R. F. (2015). Transdermal Drug Delivery: Innovative Pharmaceutical Developments Based on Disruption of the Barrier Properties of the stratum corneum. Pharmaceutics, 7(4), 438–470. https://doi.org/10.3390/pharmaceutics7040438

Alshobaili, H. A., Shahzad, M., Al-Marshood, A., Khalil, A., Settin, A., & Barrimah, I. (2010). Genetic background of psoriasis. International journal of health sciences, 4(1), 23–29.

Clonidine (transdermal). clonidine (transdermal) | Michigan Medicine. (n.d.). Retrieved September 11, 2021, from https://www.uofmhealth.org/health-library/d00044t1.

de Koning, H. D., Rodijk-Olthuis, D., van Vlijmen-Willems, I. M., Joosten, L. A., Netea, M. G., Schalkwijk, J., & Zeeuwen, P. L. (2010). A comprehensive analysis of pattern recognition receptors in normal and inflamed human epidermis: upregulation of dectin-1 in psoriasis. The Journal of investigative dermatology, 130(11), 2611–2620. https://doi.org/10.1038/jid.2010.196

Félix Garza, Z. C., Liebmann, J., Born, M., Hilbers, P. A., & van Riel, N. A. (2017). A dynamic model for prediction of PSORIASIS management by blue LIGHT IRRADIATION. Frontiers in Physiology, 8. https://doi.org/10.3389/fphys.2017.00028

Juilland M. & Thome M. Holding All the CARDs: How MALT1 Controls CARMA/CARD-Dependent Signaling. Front Immunol. 2018;9:1927. Published 2018 Aug 30. doi:10.3389/fimmu.2018.01927

Kamiya, K., Kishimoto, M., Sugai, J., Komine, M., & Ohtsuki, M. (2019). Risk Factors for the Development of Psoriasis. International journal of molecular sciences, 20(18), 4347. https://doi.org/10.3390/ijms20184347

Krappmann, D., Nagel , D., Schlauderer , F., Lammens , K., Hopfner , K.-P., Chrusciel , R. A., & Kling, D. L. (2017). (S)-Enantiomer of Mepazine. (U.S. Patent No. 9,718,811 B2).

Martin, G., Young, M., & Aldredge, L. (2019). Recommendations for Initiating Systemic Therapy in Patients with Psoriasis. The Journal of clinical and aesthetic dermatology, 12(4), 13–26.

Mayo Foundation for Medical Education and Research. (2021, September 1). Estradiol (Transdermal route) proper use. Mayo Clinic. Retrieved September 11, 2021, from https://www.mayoclinic.org/drugs-supplements/estradiol-transdermal-route/proper-use/drg-20075306.

McGuire, C., Elton, L., Wieghofer, P., Staal, J., Voet, S., Demeyer, A., Nagel, D., Krappmann, D., Prinz, M., Beyaert, R., & van Loo, G. (2014). Pharmacological inhibition of MALT1 protease activity protects mice in a mouse model of multiple sclerosis. Journal of neuroinflammation, 11, 124. https://doi.org/10.1186/1742-2094-11-124

Meloni, L., Verstrepen, L., Kreike, M., Staal, J., Driege, Y., Afonina, I. S., & Beyaert, R. (2018). Mepazine Inhibits RANK-Induced Osteoclastogenesis Independent of Its MALT1 Inhibitory Function. Molecules (Basel, Switzerland), 23(12), 3144. https://doi.org/10.3390/molecules23123144

Pastore, M. N., Kalia, Y. N., Horstmann, M., & Roberts, M. S. (2015). Transdermal patches: history, development and pharmacology. British journal of pharmacology, 172(9), 2179–2209. https://doi.org/10.1111/bph.13059

Schadler, E. D., Ortel, B., & Mehlis, S. L. (2019). Biologics for the primary care physician: Review and treatment of psoriasis. Disease-a-month : DM, 65(3), 51–90. https://doi.org/10.1016/j.disamonth.2018.06.001

Uva, L., Miguel, D., Pinheiro, C., Antunes, J., Cruz, D., Ferreira, J., & Filipe, P. (2012). Mechanisms of action of topical corticosteroids in psoriasis. International journal of endocrinology, 2012, 561018. https://doi.org/10.1155/2012/561018

van de Kerkhof, P. C., Kragballe, K., Segaert, S., Lebwohl, M., & International Psoriasis Council (2011). Factors impacting the combination of topical corticosteroid therapies for psoriasis: perspectives from the International Psoriasis Council. Journal of the European Academy of Dermatology and Venereology : JEADV, 25(10), 1130–1139. https://doi.org/10.1111/j.1468-3083.2011.04113.x

Van Nuffel, E., Schmitt, A., Afonina, I.S., Schulze-Osthoff, K., Beyaert, R., & Hailfinger, S. CARD14-Mediated Activation of Paracaspase MALT1 in Keratinocytes: Implications for Psoriasis. J Invest Dermatol. 2017;137(3):569-575. doi:10.1016/j.jid.2016.09.031.

Zhang, P., & Wu, M. X. (2018). A clinical review of phototherapy for psoriasis. Lasers in medical science, 33(1), 173–180. https://doi.org/10.1007/s10103-017-2360-1

Table and Figures

Table 1

Chemical properties of mepazine acetate

Compound Name	Mepazine Acetate
Molecular Formula	$C_{21}H_{26}N_2O_2S$
Molecular Weight	370.5 g/mol
Melting Point	187.5 °C

Figure 1

COBWEB model split into three sections to represent the layers of the epidermis

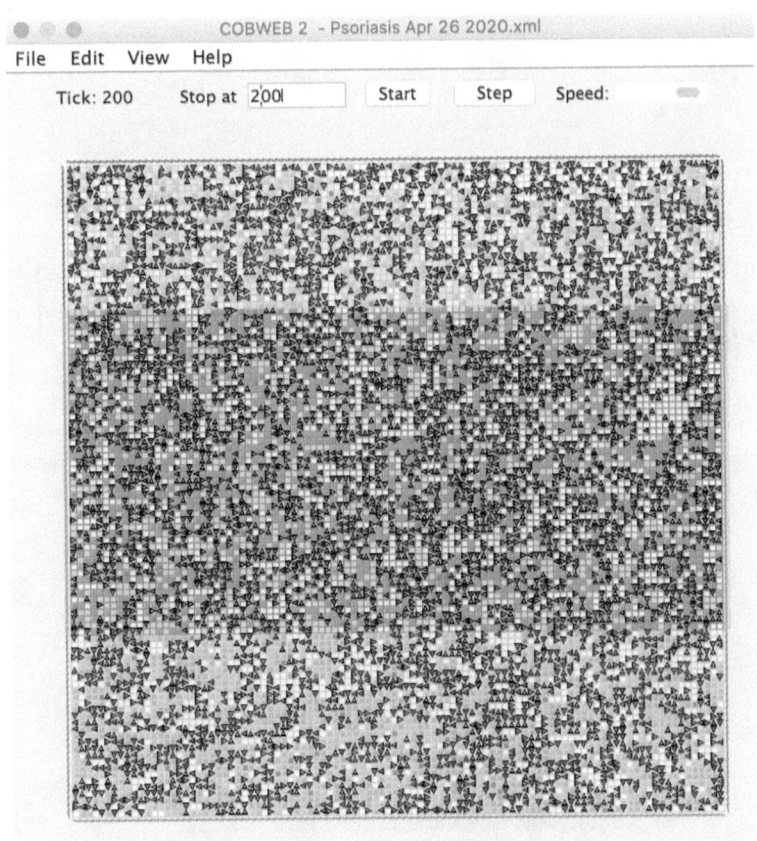

Figure 2

The average amount of CBM complex-producing dead keratinocytes on the epidermis over time between the treatment and control conditions

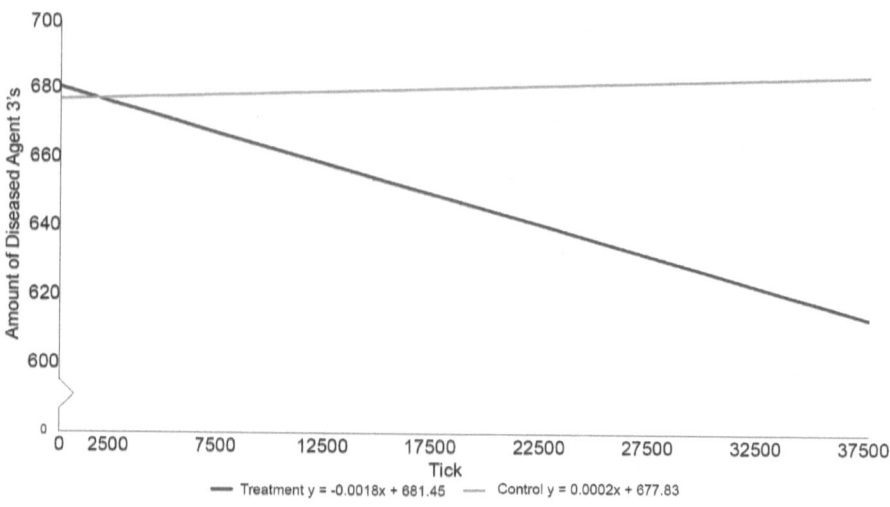

Figure 3

The mean amount of CBM complex-producing dead keratinocytes (± SE) in the stratum corneum after 35,000 ticks between models with and without treatment

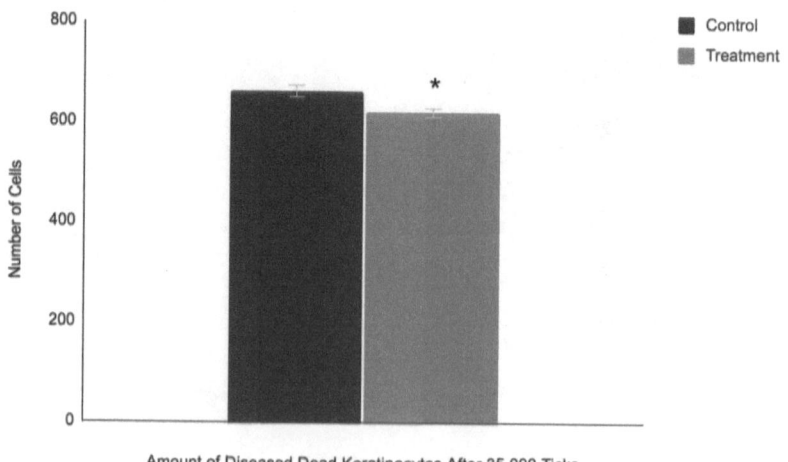

Note. The mean followed by an asterisk is significantly different (p<0.05) according to a two-sample t-test assuming equal variances.

PANEL 2: FACING NEW CHALLENGES IN THE SCIENCES

Digital Divide, Virtual Care, and Determinants of Health in a Post-COVID-19 Era

Peter Anto Johnson (University of Alberta),
John Christy Johnson (University of Alberta)

The rise and transition to virtual care has permeated both primary care and specialist healthcare settings amidst our new reality of the pandemic. For many patients and physicians who are used to conventional care, tele-healthcare has created a rift in the way they approach consults, referrals, and treatment. Inevitably however, both patients and care providers recognize virtual care has now become the new norm we must all adapt to.

Technology enabling virtual care is a valuable instrument that prevents the incidence of more variants, strains, and infectious diseases, while enabling patient care and follow-up. Moreover, advances in remote monitoring, contract tracing, and patient safety have reduced the risk of viral transmission and offer increased convenience of visits, reducing travel and wait times – one of the biggest critiques of modern primary care (Li et al., 2020). It might perhaps also allow physicians and other care providers with increased time in their schedule to see more patients by eliminating the need for shifting locations. Studies have also suggested that virtual connectivity and communication provides an environment that is less restrictive and lowers stress levels (Nissen & Lindhardt, 2017; Samargandy et al., 2020).

However, it is evident that virtual care has inherent limitations associated with the lack of in-person consults. One aspect to recognize is the fact that patients are not given the same level of encouragement or human elements of interaction (Glauser, 2020). Face-to-face communication is just not the same when doctors and patients are sitting in front of their computers rather than each other. This is because the presence of another human being in the environment can ultimately influence sensitivity of counsel where care has psychological components. Furthermore, though many novel innovations in telemedicine and remote monitoring technology are evolving, physical exams and investigations have nonetheless become more difficult in a virtual care environment (Johnson & Johnson, 2021; Mills et al., 2020; Perrone et al., 2020). Self-reports and other subjective measures can be biased without the presence of a white coat figure whose clinical expertise of physical findings are what guides a diagnosis, investigations, and medical interventions. Additionally, the dependence on technologies is another challenge, which has threatened fundamental rights of patient confidentiality, which is difficult to protect online (Hall & Mcgraw, 2014).

Moreover, despite an ease of access to care, this new norm may not be as efficient as we had first believed to be in addressing barriers that create disparity among patients. Taken together, several studies have demonstrated parallels to in-person healthcare access as a recurring theme, determining systemic racial factors, socioeconomic inequities, and limitations in rural access to care (Darrat et al., 2021; Jaffe et al., 2020; Zhai, 2020). Of these studies, certain ones also determined the influence of generational gaps in digital literacy was a poor predictor of access to virtual care (Jaffe et al., 2020; Zhai, 2020), and other studies suggested geographical proximity was more closely associated with access to care (Gajarawala & Pelkowski, 2021; Makhni et al., 2020). A body of literature, which has focused on economic status, reported patients with a lower median household income had decreased access to tele-health compared to middle- and high-income patients accessing telemedicine (Agarwal et al., 2010; Darrat et al., 2021; Harst et al., 2020). Of the studies describing demographic factors, increasing age and male sex was associated with lower odds of accessing virtual care. It was also noted Indigenous, Asian, non-English speaking, and other ethnic minority groups were less likely to use virtual services as observed by trends in in-person healthcare (Jaffe et al., 2020; Li et al., 2020; Zhai, 2020). These factors are undeniably complex and reflected structural systems, societal frameworks, and personal circumstances that access to virtual care did not address. As such, although the use of telemedicine and virtual care during the pandemic has increased the accessibility of health services to the larger population, studies have suggested a sustained disparity among already vulnerable groups - perhaps owing to inherent systemic flaws already present in healthcare.

Already recognized as here to stay, virtual care presents new opportunities along with novel challenges as well as similar limitations, in terms of barriers to access as conventional care. Regardless, as it is the new norm, it is a new reality that we must gradually start to familiarize ourselves with.

References

Agarwal, R., Gao, G. G., DesRoches, C., & Jha, A. K. (2010). The digital transformation of healthcare: Current status and the road ahead. In Information Systems Research (Vol. 21, Issue 4, pp. 796–809). INFORMS Inst.for Operations Res.and the Management Sciences. https://doi.org/10.1287/isre.1100.0327

Darrat, I., Tam, S., Boulis, M., & Williams, A. M. (2021). Socioeconomic Disparities in Patient Use of Telehealth during the Coronavirus Disease 2019 Surge. JAMA Otolaryngology - Head and Neck Surgery, 147(3), 287–295. https://doi.org/10.1001/jamaoto.2020.5161

Gajarawala, S. N., & Pelkowski, J. N. (2021). Telehealth Benefits and Barriers. Journal for Nurse Practitioners, 17(2), 218–221. https://doi.org/10.1016/j.nurpra.2020.09.013

Glauser, W. (2020). Virtual care is here to stay, but major challenges remain. CMAJ : Canadian Medical Association Journal = Journal de l'Association Medicale Canadienne, 192(30), E868–E869. https://doi.org/10.1503/cmaj.1095884

Hall, J. L., & Mcgraw, D. (2014). For telehealth to succeed, privacy and security risks must be identified and addressed. Health Affairs, 33(2), 216–221. https://doi.org/10.1377/hlthaff.2013.0997

Harst, L., Timpel, P., Otto, L., Richter, P., Wollschlaeger, B., Winkler, K., & Schlieter, H. (2020). Identifying barriers in telemedicine-supported integrated care research: scoping reviews and qualitative content analysis. Journal of Public Health (Germany), 28(5), 583–594. https://doi.org/10.1007/s10389-019-01065-5

Jaffe, D. H., Lee, L., Huynh, S., & Haskell, T. P. (2020). Health Inequalities in the Use of Telehealth in the United States in the Lens of COVID-19. Population Health Management, 23(5), 368–377. https://doi.org/10.1089/pop.2020.0186

Johnson, P. A., & Johnson, J. C. (2021). Considerations for robotic-assisted laparoscopic surgery in children. Journal of Minimal Access Surgery, 17(2), 276. https://doi.org/10.4103/jmas.jmas_327_20

Li, H. L., Chan, Y. C., Huang, J. X., & Cheng, S. W. (2020). Pilot Study Using Telemedicine Video Consultation for Vascular Patients' Care During the COVID-19 Period. Annals of Vascular Surgery, 68, 76–82. https://doi.org/10.1016/j.avsg.2020.06.023

Makhni, M. C., Riew, G. J., & Sumathipala, M. G. (2020). Telemedicine in Orthopaedic Surgery: Challenges and Opportunities. The Journal of Bone and Joint Surgery. American Volume, 102(13), 1109–1115. https://doi.org/10.2106/JBJS.20.00452

Mills, E. C., Savage, E., Lieder, J., & Chiu, E. S. (2020). Telemedicine and the COVID-19 Pandemic: Are We Ready to Go Live? Advances in Skin and Wound Care, 33(8), 410–417. https://doi.org/10.1097/01.ASW.0000669916.01793.93

Nissen, L., & Lindhardt, T. (2017). A qualitative study of COPD-patients' experience of a telemedicine intervention. International Journal of Medical Informatics, 107, 11–17. https://doi.org/10.1016/j.ijmedinf.2017.08.004

Perrone, G., Zerbo, S., Bilotta, C., Malta, G., & Argo, A. (2020). Telemedicine during Covid-19 pandemic: Advantage or critical issue? The Medico-Legal Journal, 88(2), 76–77. https://doi.org/10.1177/0025817220926926

Samargandy, S. A., Al Garni, T. A., Almoghairi, A., Alahmari, M., Alshehri, B., Mosaad, M., Nour Ahmed, J. M., Amri, H. A., & Samargandy, S. (2020). Effect of COVID-19 pandemic on the cardiac outpatients' perception of seeking medical advice. Journal of the Saudi Heart Association, 32(3), 377–382. https://doi.org/10.37616/2212-5043.1094

Zhai, Y. (2020). A Call for Addressing Barriers to Telemedicine: Health Disparities during the COVID-19 Pandemic. In Psychotherapy and Psychosomatics (Vol. 90, Issue 1, pp. 64–66). S. Karger AG. https://doi.org/10.1159/000509000

PANEL 3: RESEARCH IN MENTAL HEALTH

Mental Health Amid COVID

Belinda Tam

Antarctic Institute of Canada

Abstract

The topic and discussion of mental health and its impact on the everyday lives of individuals have been increasingly important as COVID has progressed for the past 19 months and counting. No matter the situation though, individuals, groups, and organizations can support one another as different parts of the world start to transition out of COVID and/or into other situations. This article will examine how we can not only emotionally support ourselves but each other through self-care and the discussion of the effects of procrastination and peer pressure on mental health.

Keywords: mental health, self care, emotional regulation, procrastination, peer pressure

Self Care

The term self-care was first used in the context of nursing but has now seemingly expanded to be used in multiple contexts including (but not limited to) personal care, the workforce, and academia. With that being said, this section will discuss how the concept of self-care came to be, the degrees of self-care, and how this information is of concern during the pandemic, before and possibly after as well.

In 1956, Orem "first used the idea of self-care ... [to] express the view of human beings as attending to and dealing with themselves. This view is important and complex because the individual, the self, is both the agent of action and the object of action" (Denyes et al., 2001). Today, this definition still applies to individuals and their everyday lives, especially during the pandemic as the physical well-being of an individual is placed at risk when not taking precautions such as getting the vaccine and wearing face masks. Moreover, as the body is an engine for the mind, taking care of our physical well-being is of utmost importance before considering mental health. With that being said, as the definition of self-care evolved, Levin & Idler (1983) used the term self-care in health to refer to activities "individuals undertake in promoting their own health, preventing their own disease, limiting their own illness, and restoring their own health." This is the form of mental health that we are seeing today – individuals doing activities on a larger scale such as spa day, time in nature and away from technology to the seemingly more minor habits that we build into our everyday lives by taking care of our own hygiene, getting enough sleep, and having a clean environment to work in. These self-care activities and many others help to restore the physical well-being of a person but by doing so, it also helps to restore their mental health as well. This is especially important during the pandemic as we were/are socially distancing and staying at home. As we transition back into a world that emulates a pre-pandemic version, it is also important to emphasize that many individuals are doing these activities in a private setting so in terms of academia, "it has gone unnoticed by most students ... and public health alike [but t]here is substantial and growing self-care literature for individuals to look at" (Levin & Idler, 1983). Therefore, raising awareness and getting a call from action from here on out is key.

The Degrees of Self Care

Depending on the situation that an individual finds themselves in, regardless of whether it is before, during, or after (a pandemic), the degree to which self-care is needed will vary. In the past, research has put a focus on using self-care in preventing burnout and maintaining the overall psychological wellness of an individual (Lev & Owen, 1996). However, it seems that more people pre-COVID and now are feeling burnout and distress much more regularly, even with the various advantages that may come in the present work environments.

Before discussing what can be done to mitigate this, it is important to define what distress and burnout are in terms of self-care. As described by Lev & Owen (1996), "distress is a natural state that cannot be avoided ... [and] is typically described as a subjective emotional state or reaction experienced by an individual in response to ongoing stressors, challenges, conflicts, and demands" (Lev & Owen, 1996). Burnout, on the other hand, is "characterized by feelings of depersonalization, emotional exhaustion, and a lack of feelings of satisfaction and accomplishment, and it may result from prolonged work with emotional challenging clients" (Lev & Owen, 1996). With that being said, it can be said that distress is possibly a factor that leads to burnout, and it is clear that a person having emotional regulation is key. Even when a person's social circle is supportive, if the person themselves does not know themselves well enough, the social circle may not necessarily do what is best for the individual, possibly having more negative effects on the individual (even if they do not intend it). Furthermore, a person knowing their own triggers, boundaries, etc. helps them in not only determining what their principles, values, and goals are but having this knowledge can provide the intrinsic motivation that seemingly delays procrastination (to be later discussed in this paper) and prevent impairment. "Impairment, or impaired professional competence, may refer to the deleterious impact of distress, left untreated over time... [However, d]istress does not necessarily lead to impairment, but [instead,] a lack of adequate attention to distress makes this possibility more likely" (Lev & Owen, 1996). Therefore, having emotional regulation and being in an ideal situation and/or environment will help the individual's mental health but no matter the situation, the individual should still take time to take care of themselves and the environment that they are in every day. Application of this can be found above with the aforementioned activities but can also include more "basic" things such as exercising, sleeping, and having a proper routine that is being performed on a daily, weekly, monthly, and yearly basis.

Procrastination

As a problem that not only affects students but the everyday adult, procrastination is mainly known as a "phenomenon that often entails negative outcomes [concerning] performance and subjective well-being" (Klingsieck, 2013). Especially with the pandemic, this definition seems to be highly appropriate. However, it has also been proposed "that not all procrastination behaviour [are] either harmful or lead to negative consequences" (Chun & Choi, 2005). This idea leads procrastination to have two main categories – passive and active and three types of people – procrastinators (passive procrastinators), active procrastinators, and non-procrastinators.

According to Chun & Choi (2005), passive procrastination can be defined by the traditional definition of procrastination while active procrastination can be seen as the positive procrastinator. Besides these two categories, the paper goes on to define non-procrastinators as individuals who do not procrastinate as they are "constantly engage[d] in planning and organizing … [leading to] more realistic perceptions [and control] of time" (Chun & Choi, 2005). In terms of comparing active and passive procrastinators then, the main distinction between the two is that passive procrastinators tend to have decision-making problems, not complete tasks on time, and possibly give up the task as their self-doubt feelings increase as they procrastinate. As a whole, they also "underestimate the overall time that [is] required to complete tasks" (Chun & Choi, 2005). In comparison, active procrastinators are individuals who prefer to work under pressure, "and they make deliberate decisions to procrastinate" (Chun & Choi, 2005). Furthermore, unlike passive procrastinators who have a higher chance of not completing tasks because of their feelings, "[a]ctive procrastinators are persistent and able to complete tasks at the last minute" (Chun & Choi, 2005). With that being said, one conclusion that the study came to is that "…although active procrastinators procrastinate to the same degree as passive procrastinators, they are more similar to [non-procrastinators] than to passive procrastinators in terms of purposive use of time, control of time, self-efficacy belief, coping styles, and outcomes including academic performance" (Chun & Choi, 2005). This demonstrates that the root of procrastination may come from an individual having emotional regulation problems which leads us into discussing the implications of procrastination on mental health and academic performance through self-efficacy, motivation, and stress-coping strategies.

Self-efficacy "refers to the belief that one can reliably perform the tasks that are required for successful goal achievement" (Chun & Choi, 2005).

In conjunction with motivation, "…a force that drives a person to engage in a particular activity", a person may not procrastinate. However, there are also two forms of motivation – intrinsic and extrinsic. Intrinsic motivation refers to a person's internal drive while "[e]xtrinsic motivation … refers to motivation that results from either positive or negative external contingencies" (Chun & Choi, 2005). Although both types of procrastination are negatively correlated with academic procrastination, it was demonstrated that "extrinsic motivation was necessary to prevent task delay" and intrinsic motivation led to more time devoted to the task at hand (Chun & Choi, 2005). Furthermore, as previously mentioned, due to the drive intrinsic motivation implies, it may be the only thing that delays and prevents procrastination as a whole (Chun & Choi, 2005). With that being said, emotional regulation plays into both forms of procrastination as one must have control to a certain extent or else the task at hand will not get done no matter what. This fact leads to the reasoning behind why students may procrastinate even more when studying from home during COVID as there is no change in environment and they do not have the internal drive to be able to. get their tasks done. The root cause of this may lead back to students not having higher aspirations for themselves or just plain laziness. However, more research needs to be done in order to verify how intrinsic motivation can be applied in our everyday lives, leaving us feeling more fulfilled and aiming higher.

Academic Procrastination & Stress-Coping Strategies

Many individuals are used to hearing the word procrastinate in the 21st century as many grew up being students. However, the routines and habits that we build throughout our time at school tend to affect us as we progress through life. With that being said, this can be proven through the fact that "[t]he prevalence rates of academic procrastination (i.e., procrastination of study-related activities; e.g., writing a term paper, studying for an examination) among university students are considerably [high]…" (Klingsieck, 2013). Furthermore, "[s]tudents who procrastinated were more anxious and stressed across the entire semester, and were more agitated before a test" (Klingsieck, 2013). Nonetheless, even though these statistics demonstrate the large effects of procrastination on students, this section aims to show that there are strategies when dealing with procrastination by demonstrating various perspectives and how getting oneself to the state of intrinsic motivation may be the only way to not procrastinate.

Firstly, "the differential psychology perspective understands procrastination as a personality trait" so similar to this paper, they are studying and showing the relationship between procrastination with other factors such as mental health (Klingsieck, 2013). Next, there is the clinical psychology perspective which correlates procrastination with negative results such as depression "…to determine whether procrastination is clinically relevant or not" (Klingsieck, 2013). Lastly, there is arousal procrastination which leaves the individual to believe that they perform best under pressure even though that may not necessarily be the case (Klingsieck, 2013). With that being said, these are only a couple perspectives out of many others and although these play roles in both active and passive procrastination, "[i]f it turns out that procrastination is a phenomenon with cross-situational and cross-contextual stability, the notion of procrastination as a personality variable will be supported. If it turns out that procrastination displays itself very differently in different contexts (e.g., domains), the notion of procrastination as a domain-specific phenomenon might be more appropriate" (Klingsieck, 2013). Although this area of research still needs to be further developed, it is possible that factors such as a person's principles, workload, and environment can play a role in how they react. Moreover, for a person to understand the concepts and identify which one they have can be key in determining next steps to take when mitigating stress (for both students and working adults) and stress-coping strategies for future purposes (i.e. during COVID) (Chun & Choi, 2005; O'Donoghue & Rabin, 2001). This leads us to our discussion of the logic of procrastination and the steps to take.

Understanding these concepts and applying them to one's life can be very difficult. As "[t]he logic of procrastination says that a person procrastinates because she perceives the cost of delay to be small; if the person is busy, she sees the cost of delaying as higher, and is, therefore, less likely to procrastinate" (O'Donoghue & Rabin, 2001). This concept can have larger implications attached to it as well – whether a person achieves their goals or not as "…people may procrastinate more in pursuit of important goals than unimportant ones, or equivalently that increasing importance can exacerbate procrastination" (O'Donoghue & Rabin, 2001). This brings us to what seems to be the best strategy currently – the individual should find an intrinsic motivator. When having something that motivates you, you not only find how to complete the task but it also removes any stress and discomfort caused by the task at hand. Furthermore, it may also lead to the student/individual setting proper routines, feeling more fulfilled, and having healthier habits. The passive procrastinator may eventually become a non-procrastinator and by becoming one, they will have "higher levels of purposive use of time, time control, … self-efficacy[, …] and experience

[more] positive outcomes" (Chun & Choi, 2005). This strategy may be the ideal in terms of getting more done and living a life that the individual is satisfied with, but they may prefer active procrastination which yields similar results to non-procrastination. If not, it may "actually [enhance] the well-being and [overall] performance ..." for individuals working in a "demanding, unpredictable, and fast-changing environment" (Chun & Choi, 2005). Other strategies that can be applied include building up a habit to get things done, resolving internal conflicts, or finding an autonomous reason. However, these are areas to consider when conducting future research and for future application so regardless of what a person chooses to do, they should do it while keeping their well-being in mind.

Peer Pressure

When one thinks about peer pressure, it usually comes down to someone pressuring someone to do a task that they are usually uncomfortable with. But of course, whether or not the person doing the task is actually uncomfortable depends on the person and perhaps the situation. With that being said, peer pressure may actually be very beneficial if your peers or colleagues are encouraging you to do something for your own benefit. For students and those in the workplace, this may have decreased with COVID as many things are being conducted online but this may not be the case for those working in person or in business. During the earlier part of COVID, since companies and businesses are laying people off and the workload on those who remained increase significantly, it is key to have more emotional support in the workplace and at home. This section aims to demonstrate both the positive and negative side of peer pressure and its implications on mental health (especially during COVID).

Firstly, in order to have any form of peer pressure, two components must be in place to create motivation. The first component is that an individual's effort "must affect the well-being" of another individual, if not, multiple individuals, and the second, is that the other individual can do the same as "equilibrium effort is higher" when there is peer pressure in comparison to when there is none (Kandel & Lazear, 1992). This comes with some negative consequences though – "[w]hile pressure guarantees higher effort, it does not guarantee higher utility because the pressure itself is a cost borne by all members" in the interaction and the individuals involved may not necessarily enjoy/feel comfortable working in an environment that exerts peer pressure (Kandel & Lazear, 1992).

Furthermore, pressure can be classified as internal or external. But for the purposes of this portion, this will not be discussed in detail except for stating that peer pressure is external. In relation to people in the workforce, those with a heavy workload may be facing more internal pressure than outward but managers and upper management may be putting more pressure on them. However, this cannot be called peer pressure as the employee may not or cannot do the same thing. With that being said, the amount of peer pressure that is exerted on students is high, especially those who are taking on experimental or classes with heavy collaboration portions. Moreover, in terms of incorporating emotional support in the workforce and at home then, peer circles and family members could perform tasks such as asking what they can do to help, finding ways to lighten the load, and generally do what is best for the individual. However, they should not assume what to do, but rather ask the person what they can do as you can never be sure about how your actions will affect another person if you do not ask.

Conclusion

In conclusion, the topic and discussion of mental health is ever-growing and whether we are in the middle of a pandemic or not, individuals should be taking care of themselves physically and mentally. This article discussed activities that can be done to do so, methods that can be used when dealing with procrastination and peer pressure, but as a whole, learning how to self-regulate and putting systems in place that focus on self-care and mental health are one, if not, the key factor in which an individual will live their ideal lives.

References

Chun Chu, A. H., & Choi, J. N. (2005). Rethinking procrastination: Positive effects of" active" procrastination behavior on attitudes and performance. The Journal of social psychology, 145(3), 245-264.

Denyes, M. J., Orem, D. E., & Bekel, G. (2001). Self-care: A foundational science. Nursing science quarterly, 14(1), 48-54.

Kandel, E., & Lazear, E. P. (1992). Peer pressure and partnerships. Journal of political Economy, 100(4), 801-817.

Klingsieck, K. B. (2013). Procrastination. European Psychologist.

Lev, E. L., & Owen, S. V. (1996). A measure of self-care self-efficacy. Research in nursing & health, 19(5), 421-429.

Levin, L. S., & Idler, E. L. (1983). Self-care in health. Annual review of public health, 4(1), 181-201.

O'Donoghue, T., & Rabin, M. (2001). Choice and procrastination. The Quarterly Journal of Economics, 116(1), 121-160.

Senecal, C., Koestner, R., & Vallerand, R. J. (1995). Self-regulation and academic procrastination. The journal of social psychology, 135(5), 607-619.

PANEL 4: LITERATURE, FOLKLORE, AND MYTH

Marlowe's *Faustus*, The Greatest Sin of Despair, and Lack of Repentance: Why Salvation is Impossible for Faustus

Christina Dinh Nguyen

Department of English, York University

Abstract

The issue at hand is how the religious mores at the time Christopher Marlowe wrote *Dr. Faustus* affected the presentation of the sin of despair. This requires thorough understanding of Renaissance values, and how they affected the consequent literature. Many other scholars have determined that despair is *a* central sin in this play, but I ask, why is it *the greatest* of all of Faustus's sins? Why can all his grievous sins (including, but not limited to, the denial of God and of hell, turning to illicit knowledge, and lechery) be forgiven *except* despair? What does this mean in the religious context of Marlowe's time, and how does that affect what was written? Finally, how does this affect what that audience saw, and what we see now?

We look to the Christian notion of sin as a way to understand life's ending and the afterlife. It is a useful way to view Faustus's behaviour simply because it is the only criteria which determines where he goes in the afterlife. It is important to identify the *greatest* sin, rather than several sins, because all sins can be forgiven (Matthew 12:30-32). But still the play ends unhappily and Faustus is unforgiven. Why? By looking at the greatest sin, which is presumably the sin that causes the unhappy ending, we can isolate the causes of damnation.

I find that the central sin of despair is the only unforgivable sin that Faustus commits, which damns him to hell without salvation. This further implies that there *is* an unforgivable sin; i.e. a sin that cannot be redeemed through the classification of venial nor mortal sin. To despair is to thus have, repeatedly, no faith in God's promise for salvation, in the macabre self-dooming way that Faustus sentences himself to failure. I look at the traces of the morality play in order to contextualize how Faustus's suns are presented; then I study despair specifically.

Keywords: despair, Christian sin, renaissance, damnation, medieval, temptation

Introduction

Historical Context

Renaissance humanism was the revived belief in the power of the individual. This idea had been proposed by Desiderius Erasmus and Thomas More, both of whom were early humanists. Throughout the middle ages, the understanding of Man's role in the universe was one of subservience; everyone was expected to live their lives to worship God, and to earn their way to heaven through the Sacraments and good works (World History Encyclopedia , 2019). By the 1300s, intellectuals began to question this purely religious view of life on Earth. They read the Greek and Latin classics to understand a new way of living. With the fall of Constantinople to the Ottomans, many of the ancient texts passed to Italy, with Greek scholars fleeing Ottoman rule (Hudson, 2019). Suddenly, a new trove of classical texts became available for consumption in Italy. Thus, out of renaissance Italy came the belief that the Greek and Latin classics, combined with Christian dogma, contained all the lessons one needed to lead a moral and effective life (History.com, 2010). The demand for lay education increased and wealthy families sent their children to these humanist schools, which operated separately from the Church (Szyliowicz, 1999). Unsurprisingly, some of the ideas rising from humanism clashed with the Church's values. Humanism's central dogma focused on individualism and secularism. But humanists did not reject Christianity altogether; rather they wanted to take the ancient ideas and use education to better understand the world while retaining Christian beliefs. Undoubtedly Marlowe would have been familiar with the four fathers of early humanism: Petrarch, Lorenzo Valla, Marsilio Ficino, and Pico della Mirandola (Makdisi). All four were voracious readers of classic works. Valla especially dedicated his adult life to using humanist education to help the Church, proving that the two were not, in fact, incompatible. He also proved that the *Donation of Constantine*, a Roman document which presumably proved Constantine's granting of land to the Church, was a forgery (The Editors of Encyclopaedia Britannica, 1998). The Latin used in the document did not linguistically match the Latin of Constantine's time. Thus he proved that tradition was bettered by humanism, at least on this one occasion.

Marlowe himself attended Corpus Christi, at Cambridge University, rather than a seminary (Riggs, 2004). This relatively humanist school, which arose out of the demand for lay education, was likely the foundation for his sympathy to humanistic thought. This paper will partially discuss humanist references (i.e. to Greco-Roman myths) as a way to garner audience sympathy for the plight of the central character. Thus, it is out of this school that Marlowe draws the numerous allusions to Greek myths, chiefly to illustrate each of Faustus's great flaws (i.e. pride, despair, lust, and the display of the Seven Deadly Sins). Describing Faust's plight in relation to a classical myth, such as Icarus, rather than to a Christian character (perhaps the harsh story of Adam, for whom temptation also led to a downfall), is a commentary on how we should view the Faust story (Prologue). It invites a humanistic lens which, while still living in a Christian framework, is also sympathetic to human weaknesses and less demanding of perfect obedience to God. Though Greek myths are not inherently always gentler in the treatment of its characters than Christian stories, by using sources from outside of Christian mythology, Marlowe invites an alternative perspective, which questions the typical harsh critique that a Christian lens induces.

Marlowe's *The Tragical History of the Life and Death of Doctor Faustus*, (or commonly referred to as simply *Dr. Faustus*) is an early modern English play that begs questions of the origins of temptation in human nature. Furthermore, it studies the outcomes of sin: how does a human being accept responsibility for their sin? How does this differ when considered from a medieval rather than a humanist perspective? I choose to focus on these two perspectives rather than any other (for example, the Greek notion of *akrasia*) to situate us in the paradigm shift occurring in theology at the time of the play's conception. The pass from the middle ages to the renaissance was a turbulent one, with aftershocks still being felt in the time of Marlowe. Indeed, Marlowe's unique approach to the original German Faustus story (*Historia von D. Johann Fausten*, with an anonymous author) brings alive the events with allusions to Biblical and classical myths, that is, the humanist approach which blends classic mythology and Christian stories. The origin of the character Faust (or, in Anglophonic translation, Faustus) likely started as a real person in Germany, a doctor by the name of Johann Georg Faust (British Library, n.d.). Because of his work as an alchemist, astrologer and magician, it is easy to understand why the *Historia von D. Johann Fausten* was written as a damning autobiography of his life. In fact, the full title of *Historia von D. Johann Fausten* in English (and credited to P.F. Gent) is *Historye of the damnable life, and deserued death of Doctor Iohn Faustus*, which clearly expressed the author's glee at Faustus' tortured life.

Central Focus

This paper will consider: in *The Tragical History of the Life and Death of Dr. Faustus*, why is the despair of God's salvation, and judgement of his own worthiness, Faustus's *greatest* sin? How does the presentation of despair affect our judgement of Faustus's culpability? What does this imply for scholars and audiences? This paper fits into the previous scholarly conversation as a way to remove focus from the other sins (lechery, greed, and so on): the only sin that matters in terms of Faustus's fate, is despair. For example, I shift away from Robert Ornstein's focus on the sin of greed in Faustus's hunger for black magic, wealth, and honour. Instead, I focus on the question of salvation in the play, why it matters, and what its boundaries are. Indeed, to question the prospect of salvation for Faustus, in regard to any *other* sin than despair, is to question God's ability, through the sacrifice of His only son, to save humans from their sins. To consider any other sin as irreconcilable lies perilously close to heresy in the Christian framework (which is how the play was meant to be viewed). This view is heretical in the Judeo-Christian ethos and philosophical view of the world, within which this play is written. So the focus should only be on despair. Furthermore, since despair is the greatest sin, that implies that salvation is possible until the very end of Faustus's life; that all actions prior to the final rejection of salvation is forgivable. In other words, the Lutheran (and thus renaissance) concept of predestination, and of no autonomy, is not applicable. Despite allusions to Greco-Roman myths, rather than Christian mythology, in an attempt to increase our sympathy with Faustus's plight, the doctrines supported by the story remain orthodox. Any elements of the renaissance are tortured and bipolar; but the evidence conclusively points to a Catholic interpretation rather than a pro-Reformation one.

Scope

Because renaissance theatre was, in part, a reaction to medieval play styles, including the morality play (among others, like the miracle play and the folk plays), I will outline the traces of the morality play that appear in *Dr. Faustus*. This informs the larger image of how sins are presented. When we understand how they are presented, then we can understand why despair rises to the forefront. It is also impossible to ignore the good and evil angel archetypes, or the personification of the Seven Deadly Sins, for example. Thus the chronological scope of this paper includes the 1400s to the end of the 1500s. I will consider *only* the English morality play, though certainly there were other types (e.g. the French morality play, like *Condemnation des banquets* by Nicolas de la Chesnaye).

There is no one Faust story; rather, there are hundreds of variations on the archetype in theatre, music, film, poetry, art, and literature. In many of the iterations, Faustus's greatest sin is *not* of abjuring God, nor his dabbling in the dark arts. Rather, it is his *repeated* abjuration of God, and of denying himself God's grace and healing. Living in an increasingly lapsed mindset where he does not need God, Faust seems to both believe that God is not there, and that the God that is there is one he does not need. This uncertainty remains unsolved by the end of the play. Indeed, the irony is that he turns to the devil, who only exists because God exists (section 6). Comparing Marlowe's Faustus story with other iterations provides a varied understanding of the many sins Faustus commits: what boundaries are there in categorizing an unforgivable sin in each play? Why are some sins displayed but not others?

Methodology

The methodology to be used in this study is as follows. First, there must be a thorough analysis of post-medieval religious mores, and what we can assume Marlowe knew of these mores. Secondly, there will be a close textual analysis of *Dr. Faustus* to pick out all relevant citations for each argument. Thirdly, the arguments will follow in logical, incremental steps from the known to the speculative.

This includes a review of the precursor to Elizabethan drama, the morality play, and its effects on Elizabethan drama. Then begins the study of Faustus's list of sins, followed by an analysis on the presentation of sin, then an identification of Faustus's *greatest* sin. Lastly there will be a discussion on autonomy, Christianity, and sin in renaissance literature. Furthermore, how do the classical allusions affect the audiences' perception of Faustus's culpability? In the conclusion, I will explain what this means for the modern scholar.

Literature Review

There is no shortage of literature on the causes and outcomes of Faustus's sins, nor is there a shortage of discussion about which of his many sins ultimately condemns him. Among the former topic, much has been written about Faustus's motivations, his psychiatric state, his disillusionment with

99

faith, and the manifestation of his immoral desires. Among the latter topic, much has been written about the devilish pact as the final sin, or despair as the final sin. What is missing from the conversation, however, is a study of the similarities and differences of this renaissance play with those of the Medieval era, with an eye to how these qualities affect the presentation of sin to the audience. This presentation of sin also has specific implications to the final sin of despair. After all, as we will see later, this is a morality play of sorts, not a mere story to entertain. There is a spiritual purpose: to save the audiences' souls by allowing them to witness what should *not* be done.

In the first grouping of literature, i.e. the group that focuses on Faustus's motivations, there are many that look at the decision to sin, and the manifestations of that decision. Robert West, for example, argues that Faustus's motivations are not honorable. It is tempting, he says, to view Faustus as a man whose motivations are sympathetic, but who had an unfortunate end. West compares Faustus with other Elizabethan characters who came to a bad end, but had sympathetic motivations. "Faustus does not come to magic grandly [i.e. for grand purposes]. In his dismissal of the allowable arts and sciences is no such nihilistic greatness as shows through Macbeth's tomorrow-and-tomorrow speech, nor any existential anguish at a universe that Faustus' analysis has exposed as hostile to man's mind and spirit" (West, 1974). Faustus acts for selfish reasons and rails against his mortality; he is not concerned with challenging the natural laws for the goodness of humanity. "What disturbs him is not mankind's plight but the fact that he is himself but a man," and he overreaches the bounds set upon him by divinity (West, 1974). But he compounds the sins of his prayer to devils and his pact with them by the further sins of presumption of both damnation and mercy. In the case of John Parker, he is less interested in the motivation than in the manifestation of the motivation. How does Faustus justify his decision? Parker argues that at the beginning of the play, Faustus had already made the decision to seek out devilish companionship; instead, the opening scene is a study of the inner thoughts of Faustus. He cycles "through various branches of knowledge, each more ambitious than the last, expressing contempt until he finally arrives at theology," all of which he bends to support his desire to sin (J. Parker, 2013).

In the second grouping of literature, i.e. the group that focuses on the condemning sin, academics have moved between the belief that the contract is the condemning sin, and that the continued despair is the condemning sin. Of the former beliefs are James MacDonald and Angus Fletcher. Of the latter beliefs are John McCloskey, Joseph Westlund, and Lily Campbell.

Many unsympathetic arguments state that his initial disillusionment with divinity and signing the devilish pact is the climatic sin. But if the devilish contract indeed is the final point, then why present us the rest of the play at all, save the ending? Why repeatedly place the Good Angel onstage, demanding that Faustus repent? Why do scholars appear before Faustus' death, asking him to pray for forgiveness? We must conclude that while the contract is the most dramatic onstage event which makes all Christian viewers' hearts leap, it is what happens after, in the lull of smaller events (like the summoning of Helen, and the conjuring of exotic grapes) that reveals to us the true flaw, which exists in Faustus' psyche: the stubborn commitment to despair. Joseph Westlund emphasizes that this stubbornness arises "because sin and damnation are far more real to Faustus than grace and salvation" (Westlund, 1963). One interesting exception to previous discourse is Gerard Cox, who instead lists three sins of equal weight: the rejection of a divinity, an obstinate devotion to dark arts, and despair. These three are sins against the Holy Ghost, he argues, and they all lead to each other, none weighing more than the other (Cox, 1973).

Furthermore, there is significant discourse on the Calvinist concept of predestination. Is Faustus doomed to be a reprobate from the beginning, or does he indeed have some agency in his own salvation? In other words, is he destined to reject God and the notion of hell because of his nature? Or is salvation and damnation still a choice until the point of death? Again, if he is predestined to failure, then why write this play as a warning to the audience? As Barbara Parker appropriately comments in a rejection of Paul Sellin's earlier argument of predestination, thus it is entirely probable that Faustus is not damned until he dies (B. L. Parker, 2011). The predestination argument is consequently sidelined.

John Parker acknowledges that Faustus is a victim of evil, being indecisive and two-faced. He posits that for "a stage villain to be shamelessly two-faced, he must show his real face—if only to some godlike watcher who sees in the dark and hears insidious thoughts whispered 'in solitude'" (J. Parker, 2013). After all, this is not a story without an observer. Marlowe is not writing simply to record a story, but to present a moral to the audience. He is hyper-aware of how he presents his characters; he grants a license to the audience to watch immoral acts unfold on stage without damning their own souls. This privilege "to discern evil at work where its victims are blinded, to bask in secret machinations whose unfolding they can witness without guilt, insofar as the pleasure of this voyeurism *supposedly* [my emphasis] arises from wanting ever so badly to see evil unmasked and punished" (J. Parker, 2013).

Campbell too states that we ought to view "Faustus as one whose fate is not determined by his *initial* sin but rather as one who until the fatal eleventh hour might have been redeemed" (Campbell, 1952). In other words, he was not truly damned throughout his sinful life, until he chose to reject salvation at the hour of his death. Campbell also acknowledges that the elements of the play (e.g. the hell-mouth ending, where Faustus is accompanied to hell by demons) that other scholars found "medieval" and out-of-place were rather "of the Reformation and that they constitute the essential dramatic unity of the play" (Campbell, 1952). We see this in Scene XIX, where "[h]ell is discovered" and the Good Angel proclaims, "The jaws of hell are open to receive thee" (XIX.15). I continue this conversation in this paper, particularly in section 4, looking at traces of medieval plays and how that affects the presentation of sin. This presentation ultimately leads to the irrefutable conclusion that despair is the greatest sin, not any of the deadly sins nor the devilish pact.

In my analysis of this problem, I support the argument that Faust, as an agent of his own soul's fate, had the greatest sin of repeated despair. My research rebuts MacDonald, Fletcher, and Sellin's research, and continues where B. Parker left off. The questions that still remain to be answered are: what characteristics are similar and different from medieval plays? How does this affect the presentation of sin, and how does it support the argument that despair is indeed the final sin, rather than predestination dooming Faustus from the start?

Traces of the Morality Play

The Morality Plays of the Middle Ages

The matters of *Dr. Faustus* were increasingly secular compared to English medieval plays; it was not focused on the story of a Biblical character. Compare it, for example, with the York Mystery Plays, which was a cycle of forty-eight distinct plays covering the whole of Christian history, from Creation to Judgement Day (Davidson, 2007). All the stories were from the Old and New Testaments; there was little deviance from the events and characters of the Christian canon. What was added to the stories was a bit of extra imagery, for stage drama. Thus Marlowe's deviating format provided a freedom to question traditional assumptions, such as the qualities of the afterlife (Mulryne and Shewring). These traditional qualities of the afterlife can be observed in the thirty-seventh play of the York Cycle, *The Harrowing of Hell*, where Jesus descends to Hell to grant salvation to souls. Images of fire, intensified with fireworks, and images of hell's gates falling down were largely popular (Davidson, 2007).

Presenting the story of a more common character, who they might not necessarily heard of in such mythic proportions, and one closer to the audience's lived experiences, brings to life the doubts and themes that the audience encountered on a daily basis, such as dealing with temptation. But the form of the play itself still echoed that of the traditional medieval morality play. The morality play, not to be confused with the cycle plays (also called mystery plays), provided the audience with a moral lesson, teaching them to become better Christians without mimicking the stories of the liturgy, as the cycle plays had (Petropoulous, 2019). Characters typically were personified virtues and vices, such as Purity, Hope, and Faith; and the stage itself had distinctive features like the hell-mouth (Beadle and Fletcher, 2008). In Marlowe, we do not have a specific reference to a hell-mouth feature at the scene of Faustus's descent into hell, but the imagery is very similar. "Ugly hell, gape not!" cries Faustus, invoking the image of a cavernous gap opening to receive his soul (XIX.189). The similarity between this hell and the hell-mouth of the cycle play places the audience's focus on the afterlife. All of the actions that Faustus takes in life has a direct effect on his afterlife, so any and all emphasis on hell is important.

Interestingly, as previously mentioned in the literature review, the image of hell in this play uniquely reinstates the mysteries of the afterlife, which Martin Luther felt the Catholic Church had removed. This quality, of course, is *not* seen in the Catholic mystery plays, which often described hell, and claimed to capture qualities of hell, such as the extremely decorative hell-mouth. In other medieval Catholic imagery, we see fire and demons as being an integral part of the understanding of hell (Schøsler 2007). What Luther does is remove all these specific qualities of hell, replacing it with a mysterious place (Beadle and Fletcher, 2008). A greater mystery is often more powerful in invoking fear and obedience than an imaginable hell. Indeed, "Faustus' wild mood-swings can be understood as a Lutheran response to the inaccessibility of death"(Beadle and Fletcher, 2008). By never describing hell directly through Mephostophilis (who is directly acquainted with hell), we only receive metaphorical glimpses into what hell *might* be. Though Faustus himself guesses at hell, asking where it is specifically, still Mephostophilis evades a clear answer, answering that "[h]ell hath no limits, not is circumscrib'd / In one self-place; but where we are is hell" (V.122-124). Upon hearing these vague answers, Faustus stubbornly declares, "I think hell's a fable," to which Mephostophilis replies, "Ay, think so still, till experience change thy mind" (V.129). This reminder that hell is real, and that consequences indeed exist, would have driven home the message for the audience that they should not need to experience hell to believe it – for Faustus has already provided the proof of what they should *not* do.

Good and Evil Angel Archetypes

One common archetype in morality plays was the presence of the Good Angel and the Evil Angel, appearing at the side of the central character, who represented all Christians in the world. These angels particularly appeared when the character faced a difficult moral decision. In *Everyman*, for example, the central character is advised by his angels on each step he takes towards heaven. The Evil Angel advises him to cling to worldly goods; the Good Angel encourages him to seek God above all. The central character was simply named Everyman, meant to represent all Christians seeking the path to heaven (Unknown, 2016).

These angels again appear at Faustus's side (Scenes V, VI, and XIX) when he is faced with soul-wracking decisions: he must *choose* to repent. The larger part of Marlowe's audience would have accepted the angels' characters at face value: the supernatural walked and breathed on the same Earth as humans. Magic was a lived experience. As Malcolm Gaskill writes, "Ordinary people's religions were practical rather than abstract, rooted in quotidian routines and hazards [; and] in most communities, certain individuals were respected (and feared) as specialists able to dispense helpful magic" (Gaskill, 2010). Demons tempted, witches performed magic, and souls were constantly at stake. Aside from the commoner's belief in Satan's direct interference in their daily lives, demonology was a legitimate science among literate circles. Indeed, King James VI of Scotland (later James I of England) himself published *Daemonologie, in forme of a dialogue* shortly after Marlowe's *Faustus*. Marlowe himself had to have understood the "religious culture that shaped his audience" in order to write with such "dramatic effect," triggering the very fears that his audience had in relation to their eternal souls (Anderson, 2012). This cultural phenomenon asks that scholars today immerse themselves in the beliefs that they study; Gaskill notes in *A brief introduction to witchcraft* that "demonology is deceptively hard to read without sensitivity to cultural context and a powerful leap of imagination" (Gaskill, 2010). It is the thin line between the titular character and the audience themselves that makes the play so powerful in its performance. Faustus' "desperate hunger for the forgiveness that everyone in Christendom, whatever their soteriological frameworks, is told they have only to ask for in faith exposes the differences between Faustus and the members of the audience as superficial or even illusory. What alarms [the audience] about Faustus, in the end, is not that he is so far from us but so close" (Anderson).

The portrayal of the two angels themselves does not appear satirical; they are genuinely fighting for Faustus's soul, guiding him in their own ways. "Repent, Faustus, and think upon thy soul," implores the Good Angel. But "God cannot pity thee," rejoins the Evil Angel; thus repentance appears futile (VI.12-13). They lack the irony or darker humour that might characterize anti-Church sentiments, as other humanist writings might, such as Erasmus's essay, *In Praise of Folly*. Throughout *In Praise of Folly*, Erasmus sheds light on the issues of self-deception and contradictions in Church doctrine through the character of Folly. Indeed, the angels of Marlowe's play provide a much-needed commentary on the moral to be learned. The angels ascertain the relationship between knowledge, power, and corruption (Prologue.20). The Evil Angel urges him to think of honour and of wealth, while the Good Angel reminds him that the necromantic books would corrupt.

Personification of the Deadly Sins

One of the most revealing scenes in the play is the presentation of the deadly sins. If the audience needed confirmation of Faustus's guilt (rather than misguided innocence), it is at this scene that we are decided. The presentation of the deadly sins in Scene VI is a warning to Faustus that he will be possessed by them through the course of his contract with Lucifer. His willful ignorance of these sins within himself, despite the absurdist troupe of characters before him, is ironic. It is also impossible not to notice that the presentation of the seven deadly sins echo a similar tendency to exhibit such vices in Christian morality plays that Marlowe would have seen in his younger years (such as in *The Castle of Perseverance*). Both Catholic and Protestant doctrines are presented in equal measure.

The largest irony of all is that the audience can clearly see Faustus is hopelessly entangled with the pleasures of the deadly sins, even just from the opening scene. The clearest deadly sins we spot are pride, greed, lust, and envy. Faustus is envious of immortal beings and the powers they have; he despairs in his mortal plight. In the opening scene, where he details his motivations, he despairs, "Yet art thou still but Faustus, and a man" (I.23). He reaches for what he cannot have, wishing to break the natural order where humans must not seek out illicit knowledge. He wishes for magic, which is forbidden to humans, and proclaims, "But his dominion that exceeds in this / Stretcheth as far as doth the mind of man: / A sound magician is a demi-god" (I.59-61). Faustus is also clearly guilty of excessive pride. Pride begins as self-respect and self-esteem, in understanding the power of the individual, but pride ends as arrogance and over-estimation of the individual capability. The overreach of power, the unwillingness to stay within the clearly-defined boundaries of humanity (i.e. to never aspire for magic) – this is what leads him astray. Greed and lust are natural companions of pride and envy. With his newfound power from the devilish contract, he demands more and more magic and illicit knowledge, and conjures up Helen to satisfy his base desires.

Furthermore, in the spirit of Protestantism, the Good Angel tells Faustus to read the Scriptures, rather than to perform acts of charity, or to perform rituals like praying the rosary and attending Mass. Martin Luther is recorded as having written, "Works are necessary for salvation but they do not cause salvation; for faith alone gives life," contradicting Christian morality of the middle ages, where works were the central path to salvation (Moe-Lobeda, 2002). On the other hand, the Evil Angel manipulates Faustus's desires by understanding his weakness for the cardinal sins. He promises worldly temptations, ("No, Faustus; think of honour and of wealth," says the Angel) and Faustus is bought (V.22).

In Faustus's interview with the seven deadly sins, which is equally a spectacle for the Elizabethan audience as it is for Faustus's own entertainment, we witness a conversation that mirrors his internal dialogue of denial and conceit. He asks no deeper questions, but presides over the presentation with glibness. He remains amused and aloof, curiously asking questions of some of the sins, but failing to recognize these characteristics in himself. He tells Mephostophilis, "[t]hat sight [of spectacle] will be as pleasant to me as paradise was to Adam the first day of his creation" (VI.108-109). A clearer example of the deadly sin of pride is never seen – here Faustus equates himself with the first man, the original man, who was created in God's image. Sadly, his convenient skill of removing references from context (e.g. *the wages of sin are death,* and failing to complete this statement) forgoes the fact that Adam, despite being in the image of God, was human, and thus committed sins. Faustus too, as a human, is condemned to commit sin. His interjections into the performance ridicule the vices themselves, and his comments reveals that he sees the performance as a simple spectacle to satisfy his visceral desires rather than as a learning experience. It is entirely likely that Mephostopheles intended this display to exhibit all the wonderful powers of hell, to seduce Faustus further in his infatuation for illicit knowledge. Faustus could have used this experience as an eye-opening experience that exposes him to his wrongful ways, but he does not. He falls into Mephostopheles' trap, utterly enthralled by the powers promised to him (e.g. conjuring). He is inherently unable to recognize these sins in himself as a result of being detached from *understanding* of what the sins are. This presentation of the sins simply demonstrates how blind Faustus is to his own flaws. Indeed, if he cannot recognize his sins, how can he repent and be saved?

The Sins of Which Faust is Guilty, and the Unorthodox Presentation of Sins

The Sins of Which Faust is Guilty

Before the identification and examination of sins, let us first define the context in which we present the sins. Prior to the Reformation, generally, all of Christian Europe was Roman Catholic; it had been so since the fourth century (Oakley, Cunningham, & Knowles, 1999). The Protestant Reformation cleaved western Christendom into two; it provided a way to oppose the views of the Catholic church while retaining Christian beliefs. The Reformation, born of humanistic values, became more of a political and social fight than a religious one. Indeed, there is little coincidence in the fact that *Faustus* is set in Wittenberg, where Martin Luther himself was sent as a monk. This is where Luther became the father of the Reformation, focusing on the idea that "[t]he just shall live through faith," rather than actions (*King James Version*, Romans 1:17). That is to say, actions do not matter – no charity work, no pilgrimage, no acts of penance would ever be enough. Why Wittenberg for Faustus? If it was sympathy for the Protestant doctrine, and a commentary on how to view Faustus, then why give Faustus a chance to repent throughout? The rest of the play speaks in favor of repentance and penance (see section 6), so it appears that the setting of Wittenberg is instead merely the best place to place a scholar so learned in theology (there being few cities in Europe with large scholastic influence).

Note that our Faustus had *some* faith. He believed in Christian principles. He knew of God's existence and his capability to save souls, but *chose* to turn his back on divinity, crying, "Divinity, adieu!" (Marlowe I.47). He sometimes believed that God could save others, but not him. In his desperation, he even proclaims, "But Faustus' offence can ne'er be pardoned: the serpent / that tempted Eve may be saved, but not Faustus" (XIX.41-42). In more ways than one it is difficult to see where Marlowe's sympathy lies in terms of Catholicism or Protestantism. His Faustus seems to understand that people's souls could only be saved through faith and the grace of God (but apparently not for himself). There is little mention of acts of charity, or the Holy Sacraments – not even Mass, confession, or even the Last Rites. These are acts of personal communion with God, i.e. receiving the body of Christ and absolution for sins committed against God. There is no denying that Faustus could especially use the latter. Instead we see him waste the last minutes of his life moaning and groaning about his terrible fate, rather than accepting the help of the scholars who pity him – and could have given him the Last Rites (which includes confession of sins). But also, nowhere does Faustus *reject* the Church or ritualistic approaches to faith (which Luther himself denied, saying that these did not increase the salvation of souls or spirituality) (Luther, 1988). This ambivalent approach, therefore, does not support either a Catholic nor a Protestant way of expressing faith or seeking redemption. Naturally, that makes it difficult for us to judge Faustus's struggle towards redemption. By which standards should we measure him? It can equally be said that he is a sympathetic character, or that he is despicable; if we only could choose one approach, the judgement would be clearer.

Putting that aside, Faustus wavers when deciding if Christian laws govern the universe. Even when Mephostophilis is in front of him, he is indecisive. He rejects the damnation and salvation of souls (Hargitai, 2018). Yet at other times he curses heaven, implying a belief in heaven, after all. But Faustus is a Christian, whether he likes it or not. If he abjures God, but keeps referring back to God, he is failing to turn his back on Christianity. And as a Christian of his time, he is supposed to live by the Nicene Creed. The Nicene Creed, as it would have been in the fifteenth and early sixteenth century, confirms belief in several doctrines (Burn, 1909). Furthermore, each Easter, Christians would reaffirm their beliefs in a rephrased Nicene Creed. Those beliefs included renouncing Satan, all his works, and his empty show (The Easter Vigil in the Holy Night, n.d.) (Apostolic Constitutions (Book VII)). We can get a closer look at how medieval masses were carried out through codices used at the time (Unknown., n.d.) That is, there was an intrinsic belief that Satan was a real being (i.e. the word 'his' implies a being) who interfered with daily human lives.

Sadly for Faustus, he has formally affirmed his beliefs by being a Christian, and his knowledge of the supernatural places him directly within their sphere of influence. Regardless of his belief system, he will still be judged by the ultimate authority of God because he is set in a literary world where that is known to be true. The audience has the dramatic irony of knowing that the Christian God is indeed real. Even when he rejects Christian theology, his soul lies in the hands of Christian supernaturalism. This too acts as a warning to the audience against atheism, or disbelief in Christian values; there is no escaping the Christian world (Iftikhar, 2014).

The Unorthodox Presentation of Sin

As previously noted, by describing Faustus's plight in relation to a classical Greco-Roman myth rather than to a Christian character, Marlowe gives us an alternative way to understand the story. It invites a humanistic lens, one sympathetic to human weaknesses and less demanding of perfect obedience to God.

Let us explore this in greater detail, with two specific examples of classical references: Icarus being the first (in the prologue), and Helen of Troy being the second (in Scene XVIII). They are, accordingly,

> Shortly he was grac'd with doctor's name,
> Excelling all, and sweetly can dispute
> In th' heavenly matters of theology;
> Till, swollen with cunning of a self-conceit,
> His waxen wings did mount above his reach
> And, melting, heavens conspir'd his overthrow
> (Prologue.17-22)

and

> Sweet Helen, make me immortal with a kiss.
> Her lips suck forth my soul: see where it flies!
> Come, Helen, come give me my soul again [...]
> I will be Paris, and for love of thee
> Instead of Troy shall Wittenberg be sacked
> (XVIII.101-108).

The Icarus myth is more sympathetic than the Adam story, and the myth of Helen of Troy is more sympathetic than the stories of the cities of Sodom and Gomorra, or the story of David and Uriah. Firstly, the Icarus story represents the pride of one man (or, at worst, two – Daedalus and Icarus), while the Adam story has more grave consequences. Icarus's behaviour only punishes himself, *not* the entirety of humanity. Faustus likewise damns only himself, rather than all humans who come after him. What is profoundly different, however, is that for Icarus he is present in a situation he does not fully comprehend for the first time, and it ends horribly. Icarus himself is escaping a prison island with Daedalus, and pushes his freedom too far. Meanwhile, Adam broke the rules with the knowledge that his Creator had forbidden it, and was severely punished as a result. On the other hand, for Faustus, there had already been 1500 years of Christianity to teach moral standards. There can be no naïveté or ignorance pleaded here. So by equating Faustus (who is incredibly guilty of willful sin) with the more pitiable Icarus, rather than someone worse (Adam, who also willfully ignored the teachings of God), the audience is inclined towards sympathy rather than outright harsh judgement.

Second, if we compare Faustus's lust for extra-marital alliances in the context of Helen of Troy, there is considerably more sympathy there than in any parallel Old Testament stories. Pick, for example, the story of Sodom and Gomorra, two cities which were entirely destroyed by the wrath of God for their immoral behaviour (Genesis 18:20). Or perhaps Marlowe might have used the story of David and Uriah to contextualize Faustus's temptation to fleshly desires. In this case, in the Euripides version of the Troy story, *The Women of Troy* and *Helen*, Helen is the victim, as are Menelaus and Paris, of the goddess Hera's jealousy. We use the Euripides' text because it is the earliest version of the Helen story, the basis for future iterations, and the version that humanists would have studied. Our Faustus, if we compare him to Paris, is a victim; we should be sympathetic. He is a victim simply of seduction and infatuation, rather than of abandoning God. Like Paris, he falls in love with an *illusion* of Helen. Faustus cries, "I will be Paris, and for love of thee / Instead of Troy shall Wittenberg be sack'd" (XVII.106-107). In other words, because of the fight over one woman's beauty, the birthplace of Protestantism is destroyed. Lust, greed, and idolatry destroy religion. Furthermore, he is enthralled by the idea not of bedding Helen – he conjures her because of a greed for power, a greed of ownership. He is pleased to own, to conjure up an illusion of such a historically weighted character. What satisfies him is not Helen herself, but the fact that he is able to produce such a powerful illusion that other men can only dream of. A Christian comparison might be hastier and harsher. "Then the Lord rained upon Sodom and upon Gomorrah brimstone and fire," does not speak of mercy or compassion; but the misfortune of Helen of Troy and Paris does (Genesis 19:24). After all, Faustus did proclaim, "Marriage is but a ceremonial toy" (V.151). This blatant rejection of Church ceremony, of sanctifying a union before God, is also a rejection of faith. By denying the need for God's consent to a marriage, Faustus restructures the relationship between man and woman, redefining the rules that God had created with Adam and Eve. So using a harsh Christian comparison, rather than a forgiving classical comparison, would not have been out of line. Faustus is lucky that we are given the references to Paris; it sways our compassion to his benefit. The other Christian story that might have been used, instead of the Troy myth, was the David and Uriah story: David takes Uriah's wife as his own in a frenzy of greed and lust (2 Samuel 12:7-15). There is no sympathetic ending; as punishment, David's first child is struck by the Lord and is made ill. Here, with the Helen story, we are tempted to forgive, because of the lightened comparison. Because Marlowe tends to use classical comparisons to evoke sympathetic responses, we know that Marlowe desires for the audience to view the character in a more sympathetic light.

Unfortunately for Faustus, this presentation in a favorable light still does not save him from a tormented fate, because he despairs and does not repent. However much we sympathize, he still committed these sins willingly, despite years of education in the Christian way. Because he lives in a Christian universe, and is bound by Christian laws (not Greco-Roman), he must answer to the Christian God. He can still be saved – if he repents. But he does not. Ultimately, despite any references to Helen of Troy, Jove, Icarus, or Penelope of Saba, Faustus's soul is measured according to the same laws that bound Sodom, Gomorra, and David. There is no escape possible, save for repentance, as evidenced in the momentous final scene. Since he refuses even to repent, through stubbornness and despair, even this final chance at salvation is gone.

The Greatest Sin

In an early scene, Mephostophilis tells Faustus that "by aspiring pride and insolence," God cast Lucifer and Mephostophilis down from heaven to suffer by being "deprived of everlasting bliss" (III.82). So Faustus really only has himself to blame for his Lucifer-esque traits of pride and insolence. In a sense, Faustus and Lucifer's sins appear to mirror each other; but Faustus is twice-damned, whereas Lucifer is once-damned. Not only is he guilty of pride and insolence, but he refuses to acknowledge that this can be forgiven. The Christian consensus is that God sent Jesus to relieve humans of their sins, and not Lucifer. Though all humans have sinned and some angels have sinned, humans can be forgiven; angels cannot. For humans, "He made Him who knew no sin to be sin on our behalf, that we might become the righteousness of God in Him" (2 Corinthians 5:21). For the angels, God "will also say to those on the left, 'Depart from Me, you cursed, into the everlasting fire prepared for the Devil and his angels" (Matthew 25:41). Lucifer as an angel is exempt from salvation – and so his pride is different from Faustus's pride. Faustus undoubtedly has been taught that his crimes have been paid for by God's sacrifice, yet he still despairs. This is not the case for Lucifer. What this implies is that even Lucifer is not a free agent; rather, his autonomy is given to him by God. Even in the traditional assumption that Lucifer is the antithesis to God, his existence is dictated by the existence of God, by the nature of his being an angel created by God. In other words, he cannot exist if God does not exist, and his powers are dictated directly as a result of his relationship with God. This reinforces the argument that Faustus is excessively proud, by attempting to subvert the natural order of 'God above all,' by seizing full autonomy, which Lucifer also tried (and failed) to subvert. All beings are given their powers (good or bad) by God; there is no other way (Macdonald, 2014).

This conversation belies arguments that Faustus's punishment is undeserved. Heroic ambition and excessive pride – Mephostophilis clearly warned Faustus of them, and Faustus did not listen. Mephostophilis is telling Faustus that he would be as damned as Mephostophilis if he consigned his soul to hell. Why warn Faustus, if as a devil, it is in Mephostophilis's best interests to ensure the contract is signed? By blurring the lines between human and demon (we are sympathetic to both), our attention is even more drawn to the fundamental difference between them. It is possible that Mephostophilis, not being able to claim salvation as a fallen angel, is showing Faustus that he is able to claim salvation from his sins, as a consequence of being human. If Lucifer, an angel, could not escape the consequences of his flaws, how could a mere mortal like Faustus hope to go unpunished for his, or at the very least, without repentance, which all humans are capable of?

Let us study this in greater detail. By presenting this play in the fashion of a morality play, where all human sins go punished if unrepented, Marlowe aligns himself with the Church. There is no heterodoxy in the doctrine as some scholars have suggested (Westlund, 1963). The method of presenting the doctrine, however, is unorthodox. What is untraditional is the unhappy ending for our main character. In *Everyman,* for example, Everyman ends in heaven after a strenuous ascension. Likewise, in *Castle of Perseverance*, Mankind receives God's mercy and is delivered from Hell. Yet Faustus does not fall into this pleasing category. At his point of death, Faustus refuses to reflect on each of his sinful actions committed over the years, and does not renounce the actions of his misspent life. In the throes of his death, he cries, "I'll burn my books! – Ah, Mephostophilis!" which tells us two things: first, it does not indicate a regret for the illicit knowledge; rather it indicates that he still does not understand the magnitude and seriousness of his sins outside of despair (XIX.190). He does not understand why they are an offense against God, and as an extension, why Lucifer might have been punished similarly. Again, I say that he is twice-damned: once for his forgivable sins, and once for the unforgivable sin of despair. This psychology, this dissonance between action and understanding of action, is likely what caused Faustus to ignore Mephostophilis's earlier warning about repeating Lucifer's crimes (see section 4.1 above). Secondly, Faustus's last cry is not a call for God and forgiveness, it is still a reach to hell. Even in the last remnants of life, he is unable to recognize his actions for what they are – he is fearful of hell and damnation, but does not believe in heaven and salvation. Indeed further proof of this can be found by looking back a little, to when the clock strikes eleven. He attempts to cry out to God, but his lack of conviction in heaven and salvation turns him to cry out for Lucifer instead. Faustus observes aloud that "[o]ne drop [of Christ's blood] would save my soul, half a drop. Ah, my Christ! / Rend not my heart for naming of my Christ; / Yet will I call on him. O, spare me, Lucifer!" (XIX.147-149).

Furthermore, as Faustus enters the last moments of his life, he wastes precious time. He spends much time debating how to escape hell:

> Mountains and hills, come, come, and fall on me,
> And hide me from the heavy wrath of God!
> No, no:
> Then will I headlong run into the earth.
> Earth, gape! O, no, it will not harbour me.
> (XIX.152-156)

and how he can save his soul, and even how he can escape suffering for eternity, in creative ways:

> O God,
> If thou wilt not have mercy on my soul,

Yet for Christ's sake, whose blood hath ransom'd me,
Impose some end to my incessant pain;
Let Faustus live in hell a thousand years,
A hundred thousand, and at last be sav'd.
(XIX.165-170)

He despairs and dangerously assumes that God will not save his soul, undervaluing God's ability to save. In all his manic desperation, he runs from one possible solution to another. But he fails to recognize the *one simple action* that can induce salvation: repentance. Because he did not review his sins, as I previously noted, he cannot repent, and cannot be saved. What he is so blind to – repentance for salvation - is what the audience recognizes easily, and longingly desires for him to see. The message for the audience is not to avoid making a pact with the devil or even committing any of the deadly sins – it is to not refuse repentance.

Conclusion

This play recognizes that according to traditional Christian doctrine, Faustus should not have been saved except through repentance. His willful ignorance makes him not a noble character undone by his despair; instead, he is the ironic scholar who was so learned in all fields except in self-examination. His failure to recognize his own dark affair with the deadly sins, compounded with his great sin of repeated despair in God's ability to save him, is what ultimately leads to his untimely demise (Escobedo). Because Marlowe merely borrows from humanist styles (such as in referencing classical myths, and reinstating the mysteries of the afterlife), and still gives Faustus the chance to be saved until death, autonomy is given; salvation is in the hands of Faustus himself. The concept of predestination (from the moment Faustus decides to reject God and sign the pact) is rejected. A close examination reveals that the similarities with Catholic thought outweigh the renaissance qualities, revealing a message firmly rooted in medieval ways, rather than the emerging renaissance.

References

Anon., and Mark Eccles. (1994). The Castell of Perseverance.

Allan, Neil. (2005). An Age in Love with Wonders: The Philosophical Context of Renaissance Literature. Literature Compass, 2(1).

Anderson, D. K. (2012). The Theatre of the Damned: Religion and the Audience in the Tragedy of Christopher Marlowe. Texas studies in literature and language, 54(1), 79-109.

Apostolic Constitutions (Book VII). (n.d.). Retrieved February 25, 2021, from New Advent: https://www.newadvent.org/fathers/07157.htm

Beadle, Richard., and Alan J. Fletcher. (2008). The Cambridge Companion to Medieval English Theatre (2nd ed.). Cambridge University Press.

British Library. (n.d.). The English Faustbuch, 1592. Retrieved April 2, 2021, from British Library: https://www.bl.uk/collection-items/the-english-faustbuch-1592#

Bryson, Michael. (2020). The Humanist (Re)Turn: Reclaiming the Self in Literature. Routledge.

Burn, A. E. (1909). Facsimiles of the creeds from early manuscripts. London: Harrison and Sons. Retrieved February 13, 2021, from Archive. org.

Campbell, L. (1952, March). Doctor Faustus: A Case of Conscience. Modern Language Association, 67(2), 219-239. Retrieved December 22, 2020

Cox, Gerard H. (1973). Marlowe's 'Doctor Faustus' and 'Sin against the Holy Ghost.' The Huntington Library Quarterly, 36(2), 119-137.

Davidson, Clifford. (2007). Festivals and Plays in Late Medieval Britain. Ashgate.

Delviccio, Dorren Alexaitdra. (1982). Theology's Tragic Glass.

Drabble, M. (1985). The Oxford Companion to English Literature (5th edition. ed.). Oxford: Oxford UP. Retrieved August 28, 2020

Dyer, J. (1999, November 19). The Medieval Mass and Its Music. Retrieved February 12, 2021, from https://www.arlima.net/the-orb/encyclop/culture/music/orbdyer.htm

Escobedo, Andrew. (2017). Volition's Face: Personification and the Will in Renaissance Literature. University of Notre Dame Press.

Fletcher, Angus. (2005). Doctor Faustus and the Lutheran Aesthetic. English Literary Renaissance, 25(2), 187-209.

Fry, S. (2007). Mythos: The Greek Myths Retold. New York City: Penguin. Retrieved December 2, 2020

Gaskill, M. (2010). Witchcraft, a very short introduction. New York City: Oxford UP. Retrieved 2020

Gates, Daniel. (2004). Unpardonable Sins: The Hazards of Performative Language in the Tragic Cases of Francesco Spiera and Doctor Faustus. Comparative Drama, 38(1), 59-81. Western Michigan University.

Goethe, J. W. (2005). Faust. (B. Taylor, Trans.) Hazleton: Pennsylvania State University. Retrieved November 3, 2020

Hargitai, Marta. (2005). From 'Resolute' to 'Dissolved': Tracking Faustus's Decision. The AnaChronist, 18(2), 243.

History.com. (2010, October 18). Italian Renaissance. Retrieved April 1, 2021, from History.com: https://www.history.com/topics/renaissance/italian-renaissance

Hudson, M. (2019, August 29). Fall of Constantinople. Retrieved March 28, 2021, from Britannica: https://www.britannica.com/event/Fall-of-Constantinople-1453

Iftikhar, Shabnum. (2014). Dr. Faustus -- a Sermon against Atheism. Language in India, 14(7), 495.

Jules Barbier, M. C. (n/a). Charles Gounod: Faust (libretto). Retrieved August 19, 2020, from http://www.impressario.ch/libretoo/libgoufau_f.htm

Kerr, T. (2009). Thomas Aquinas: a very short introduction. Oxford: Oxford UP.

Kroger, W. (2003). Lektüreschlussel: Johann Wolfgang Goethe - Faust I. Stuttgart: Reclam.

Luther, M. (1988). Large Catechism. St. Louis, Missouri: Concordia Publishing. Retrieved January 2, 2021

Macdonald, James Ross. Calvinist Theology and 'Country Divinity' in Marlowe's 'Doctor Faustus.' Studies in Philology, 111(4), 821-844.

McCloskey, John C. (1942). The Theme of Despair in Marlowe's Faustus. College English, 4(2), 110-113.

Marlowe, C. (2005). The Tragical History of the Life and Death of Doctor Faustus. New York City: Routledge. Retrieved 2020

Makdisi, George. (1990). The Rise of Humanism in Classical Islam and the Christian West: With Special Reference to Scholasticism.

Moe-Lobeda, Cynthia D. (2002). Healing a Broken World: Globalization and God.

Moore, A. (2016, September). The Infinite and the Divine. London.

Mulryne, J.R., and Shewring, Margaret. (1991). Theatre of the English and Italian Renaissance. St Martin's Press.

Oakley, F. C., Cunningham, L., & Knowles, M. D. (1999, July 26). Roman Catholicism. Retrieved April 16, 2021, from Britannica: https://www.britannica.com/topic/Roman-Catholicism

O'Brien, Margaret Ann. (1970). Christian Belief in Doctor Faustus. English Literary History, 37(1), 1-11.

Ornstein, R. (1968). Marlowe and God: The tragic theology of Dr. Faustus. PMLA, 83(5), 1378-385.

Ornstein, R. (1968, October). Marlowe and God: The Tragic Theology of Dr. Faustus. Modern Language Association, 83(5), 1378-1385. Retrieved

December 24, 2020

Parker, Barbara L. (2011). 'Cursèd Necromancy': Marlowe's Faustus as Anti-Catholic Satire. Marlowe Studies, 1(1).

Parker, J. (2013). Faustus, Confession, and the Sins of Omission. English Literary History, 80(1), 29-59.

Petropoulous, J. (2019, November 10). AP EN 2140 Drama. Medieval Drama. Toronto, Ontario, Canada. Retrieved April 2, 2021

Riggs, D. (2004). The World of Christopher Marlowe. USA: MacMillan. Retrieved February 2, 2021

Riggs, D. (2014). The World of Christopher Marlowe. London: Faber and Faber. Retrieved 2020

Rosen, B. e. (1991). Witchcraft in England, 1558-1618. Amherst: University of Massachusetts. Retrieved 2020

Schafarschik, W. (2001). Lektüreschlüssel: Johann Wolfgang Goethe - Faust II. Stuttgart: Reclam. Retrieved 2020

Schølser, Lene. (2007). Paul Géhin (Ed.), Lire Le Manuscrit Médiéval. Observer et Décrire. Zeitschrift Für Romanische Philologie, 123(4).

Sharpe, J. (2001). Witchcraft in Early Modern England. Harlow: Pearson Education.

Suárez Lafuented, María Socorro. (1996). Dos Siglos de Leyenda Faústica: Del Volksbuch Al Faust de Goethe Vía Christopher Marlowe. Archivum (Oviedo), 46(7), 451-472.

Szyliowicz, J. S. (1999, July 26). The humanistic tradition in Italy. Retrieved April 1, 2021, from Britannica: https://www.britannica.com/topic/education/The-humanistic-tradition-in-Italy#ref47542

The Easter Vigil in the Holy Night. (n.d.). Retrieved February 23, 2021, from Liturgies.net: http://www.liturgies.net/Liturgies/Catholic/roman_missal/eastervigil.htm

The Editors of Encyclopaedia Britannica. (1998, July 20). Donation of Constantine. Retrieved April 2, 2021, from Britannica: https://www.britannica.com/topic/Donation-of-Constantine#:~:text=Donation%20of%20Constantine%2C%20Latin%20Donatio,(reigned%20314%E2%80%93335)%20and

The Editors of Encyclopaedia Britannica. (2005, May 27). Priesthood of all believers. Retrieved January 23, 2021, from Britannica: https://www.britannica.com/topic/priesthood-of-all-believers

Unknown. (2016, January). Everyman. Retrieved March 28, 2021, from Coldreads: https://coldreads.files.wordpress.com/2016/01/everyman.pdf

Unknown. (n.d.). MS 74236 Missal ('The Sherborne Missal'). Retrieved February 23, 2021, from British Library Archives: http://access.bl.uk/item/viewer/ark:/81055/vdc_100104060212.0x000001?_ga=2.195633836.1863231137.1614131761-1163465495.1613570662#?c=0&m=0&s=0&cv=9&xywh=-2504%2C-1%2C10382%2C6678

Wasson, J. (1979). The Morality Play: Ancestor of Elizabethan Drama? Comparative Drama, 13(3), 210-221. Retrieved from Comparative Drama

Watt, I. (1997). Myths of Modern Individualism: Faust, Don Quixote, Robinson Crusoe. Cambridge: Cambridge University Press. Retrieved 2020

West, R. (1974). The Impatient Magic of Dr. Faustus. English Literary Renaissance, 4(2), 218-240. Retrieved 2020

Westlund, J. (1963). The Orthodox Christian Framework of Marlowe's Faustus. Studies in English Literature, 1500-1900, 191-205. Retrieved April 2, 2021

World History Encyclopedia . (2019, June 17). The Medieval Church. Retrieved April 1, 2021, from World History Encyclopedia : https://www.ancient.eu/Medieval_Church/

Percy Jackson vs Greek Mythology

Kian Isaac, Austin Mardon, and Jessica Jutras

Antarctic Institute of Canada

Author Note

This article and its research were made possible by the Antarctic Institute of Canada and its founders, Austin and Cathrine Mardon. As well as AIC Conference hosts Zach Schauer, Jessica Jutras, and Zach Irving. A special thanks to Rick Riordan for making childhoods everywhere just a little more magical.

Abstract

This article is going to cover the research that went into the Percy Jackson Franchise by author Rick Riordan. It will compare some elements of the Percy Jackson universe to a contemporary view of Greek mythology. Riordan uses a more contemporary approach in his fiction and to make sure that all of his information was correct, I also limited my searches to contemporary retellings of Greek mythology.

Keywords: Percy Jackson, Rick Riordan, Greek mythology, research, young adult fiction

I wanted to know more about the research that went into the Percy Jackson Franchise by author Rick Riordan. If you're not familiar with him or his work, the main thing that you need to know is, he writes young adult fantasy. Writing books, especially fantasy books, requires a lot of research, and I was curious about how much research went into the books. The Percy Jackson books in particular would require quite a bit of research because Riordan is taking elements from history that actually happened. If you haven't read them, these books all center around Greek mythology and while the books themselves are fiction, they use real-life inspirations, not an entire fictional world that he makes up on his own. It would be so much easier if I could just ask Rick Riordan precisely that. "How much research did you put into actual Greek Mythology when you were writing the Percy Jackson franchise?" but alas, I am not fortunate enough to be in the presence of Mr. Riordan, so I had to do my own research.

There are many versions of Greek mythology dating back to when they only used stone tablets to write on. There are also different versions of the events and creatures that came over thousands and thousands of retellings. Obviously, we can not go back and read everything from stone tablets forward and the characters that Riordan uses as inspiration are more modern, like the fact that we now know Satyrs and Centaurs are different. They used to both be classified as Satyrs, stating that they could either have the body of a horse or a goat (Apollodorus, 1998). In more modern narrations, there is a creature with the body of a horse, however, called Centaurs.

I started by rereading the first book: Percy Jackson and the Lightning Thief. I know what you may be thinking; it would be easier just to watch the movies, right? Wrong. The movies don't follow the books at all, and even Rick Riordan himself hates them, which is evident by his Twitter page. In a series of tweets with a fan, probably my favourite tweets to ever exist, Riordan states that the entire movie should be censored: "Just two hours of blank screen" (as cited in Haylock, 2020). When fans asked him why the scene with the lotus-eaters in the casino was censored on Disney's streaming service, he said, "to you guys, it's a couple hours of entertainment. To me, it's my life's work going through a meat grinder when I begged them not to do it" (as cited in Haylock, 2020). So, I think it's safe to say he was not happy with how they interpreted his story.

The main characters in the books are Percy Jackson, who is the half-blood son of Greek God Posiden. A half-blood is a child who is the offspring of a human and a God. They can also be referred to as demigods. There is Annabeth, who is the half-blood daughter of Athena, the goddess of war and battle strategy, as well as wisdom. The third person to make up their main group of friends is Grover, who is a Satyr, a creature that you will learn more about as I get into my research. There are quite a few antagonists in the book, including Luke Castellan who is the son of Hermes, who befriends Percy only to use him for his own plan to wage a war between the gods and then betray him later by framing him for stealing Zeus' lighting bolt and Hades' Helm of Darkness, which is where his power comes from. This ultimately starts the war because Hades thinks that Percy stole The Helm for Zeus and Zeus thinks Hades stole his lightning bolt. The three gods, Zeus, Poseidon and Hades, also known as 'The Big Three,' all made a pact to never reproduce with a human, Poseidon then broke that pact by having Percy, which caused some tension between them and thus an opening for Luke to create chaos in the future. Hades is the god of the Underworld, where they have to go to get to Olympus and save Percy's mom to stop the war.

The first character you meet in Lightning Thief is Grover. Grover is classified as half human and half goat, known in Greek mythology as a Satyr. Specifically, his top half is human, and his bottom half is a goat. In my own research, I came to find that Satyr's are, in fact, real. The Library of Greek Mythology, by Apollodorus, states that Satyr's are demons who attended Dionysos, the god of the grape harvest, winemaking (Apollodorus, 1998). It says they had a thick tail, and the lower half of their body was like that of a goat (Apollodorus, 1998). It also states that they are usually ithyphallic, which means that, especially of a statue or other representation of a deity, they are shown as having an erect penis. I know that part makes no appearance in Riordan's books, but I am sure it is only because that would not be an appropriate thing to put into a children's book.

Next, are The Lotus-eaters which are some of my favourite characters from Riordan's books. The scenes with them, I feel, are some of the most interesting parts of the books. The definition of what a Lotus-eater is, is similar in both portrayals of myth and the fiction of the Percy Jackson universe, just executed differently. Outside of the Percy Jackson universe, in Odyssey IX, 'Lotus-eater' denotes "a person who spends their time indulging in pleasure and luxury rather than dealing with practical concerns" (Homer, 2019). Lotus-eaters are instead, simply people living on an island that contains a tree of unknown origin (Homer, 2019). This tree has fruits and flowers named, Lotus Flowers. When people eat these foods, it causes them to be sedated, apathetic, and sleep peacefully. This primary food source was a narcotic and after they ate the lotus, they would forget their home and loved ones and only long to stay with their fellow lotus-eaters. Those who ate the plant never cared to let anyone know where they were nor return home. In Riordan's retelling, the lotus-eaters are the people that get sucked into a hotel, and the video games inside the hotel are what suck them in. Percy, Annabeth, and Grover go to the hotel in search of a place to sleep and get sucked into the games themselves. When Percy notices that something feels off, he pulls himself out of the trance, and when he finally finds his friends, they all realize that they have been stuck in the hotel unknowingly for five days. When they talk to other people there, Lotus-eaters by definition, they see that the different people have been staying at the hotel and playing those games unknowingly for multiple years as many of them are from completely different time periods.

The next character that I want to discuss is Medusa. You have probably heard of her, a powerful lady with hair of snakes that can turn people to stone. In Percy Jackson, Medusa makes her appearance as the owner of Aunty Em's Emporium. Percy and his friends don't automatically know her to be Medusa. She disguises herself with a veil to cover her face and hair of snakes. Grover ultimately is the person to uncover her real identity because he finds a statue of a satyr that looks like his uncle, and it turns out that it is. Medusa uses her eyes and hair to turn people who come into the emporium into stone.

Medusa, in Greek mythology, is described the same as in the book, with hair of snakes and a hideous face. However, she was not born that way. She was cursed and turned into a "hideous monster" because Posiden could not resist temptation, and he impregnated Medusa at the Temple of Athena (Story of the Snake-Haired Gorgon). Athena found out, and because she hated Poseidon over their battle for Athens, she cursed Medusa. It is said in every retelling of her origin story that her gaze was so piercing, the mere sight of her was sufficient to turn a man to stone. Her end came quickly when her head was severed by Perseus; which is where Percy's name comes from in Riordan's books. He, Perseus, then used her head as a weapon which still worked when it was not attached to her body before he sent her severed head to Athena.

The last thing I want to talk about in this article is not a character but rather Percy's prophecy. The prophecy is the central idea of the book; it is what guides the whole story. Percy's prophecy is as follows:

You shall go west, and face the god who has turned,

You shall find what was stolen, and see it safely returned,

You shall be betrayed by one who calls you a friend,

And you shall fail to save what matters most, in the end.

It foretells Percy's journey. "You shall go west and face the god who has turned" (Riordan, 2010, 22:51). This is when Percy, Grover, and Annabeth travel west over the United States to find the way to the Underworld. Percy thinks that it is Hades who stole the bolt to get back at Zeus for banishing him to the Underworld, turning him from 'good' to 'evil,' as it were.

"You shall find what was stolen and see it safely returned" (Riordan, 2010, 23:00). This one may be self-explanatory, but I am going to explain it anyway. This part of the prophecy refers to Zeus' lighting bolt. It is what was stolen, and it is up to Percy and his friends to see that it is returned to Zeus undamaged.

"You shall be betrayed by one who calls you a friend" (Riordan, 2010, 23:10). This one only becomes evident as you read the book. Still, you may have already guessed from the beginning of this that this line foreshadows Luke Castellan's betrayal of Percy. Luke starts by getting close to Percy, and because Percy was not friends with anyone yet, Luke being the first person he meets, he thinks they are friends. Luke gives no indication that he will turn out to be 'the bad guy,' and Percy does not even know until he realizes that the lightning bolt is in his backpack, presumably put there by Luke, which is part of Luke's plan to get Percy killed.

"And you shall fail to save what matters most, in the end" (Riordan, 2010, 23:21). This is not referring to failing his quest to return the bolt to Zeus. In fact, they succeed in that quest, which means that the lightning bolt is not what matters most. This line refers to the fact that when they go to visit Hades in the Underworld, they only have three of the pearls they need to get out of it, and subsequently have to leave Percy's mom behind because she, along with Percy, Annabeth, and Grover make four people. The story does not end horribly like you might be thinking; Hades decides to return Percy's mother after he gets his Helm of Darkness back.

All this long-winded explanation to say that I wondered if prophecies had any place in history, not just in fiction. In Greek mythology, there are people who are deemed Oracles, and these Oracles are said to be a gateway to knowing the will of the Gods, a cosmic information superhighway for understanding what lay ahead. Individuals, cities, and kings would come from across the ancient world to put their questions about their future plans to the Delphic Oracle and wait to receive a response about what the Gods thought of them. However, like the prophecies in Riordan's telling, they are often worded ambiguously and can be interpreted differently by everyone who hears it. As if that wasn't confusing enough, it is said that there were multiple Oracles, and each of them would have different answers to God's questions (History Extra, 2010).

It is clear that Riordan put a lot of research into his books and the only thing that is fiction are the fine details that make up the story, the kids, and the journey they go on. Everything he includes about Greek Mythology, the way he describes the people who lived so long ago, all has truth to it. While these books and the events that take place come from his imagination, it's safe to say that Riordan is someone you can trust for getting real-life information and events correct and the reason that these books are so timeless even though they talk about things that happened so long ago is that it gives people an escape, people see themselves in these books being Greek Gods and Goddesses. Not to mention the fact that it's a fun way to learn about history for people who aren't necessarily research inclined. That is made possible by the fact that we now know Riordan puts so much research into his books.

References

Apollodorus. Library of Greek Mythology. New York, Oxford University Press Inc, 1998.

GreekMythology.com, The Editors of Website. "Medusa: The Real Story of the Snake-Haired Gorgon". GreekMythology.com Website, 09 Jul. 2021, https://www.greekmythology.com/Myths/Creatures/Medusa/medusa.html. Accessed 16 July 2021.

Haylock, Z. (2020, June). "Percy Jackson Author Says the Movies Put His Life's Work 'Through a Meat Grinder'", Vulture. Retrieved from https://www.vulture.com/2020/06/rick-riordan-twitter-percy-jackson-movies.html

History Extra. "Ancient prophecy: oracles and the gods" HistoryExtra. com, 2010, https://www.historyextra.com/period/ancient-greece/ancient-prophecy-oracles-and-the-gods/

Homer. The Odyssey. Translated by Samuel Butler, 2019. Project Gutenberg, www.gutenberg.org/files/1727/1727-h/1727-h.htm#chap09

Riordan, R. Percy Jackson and the Lightning Thief. Narrated by Jesse Bernstein, Audible, 2010. Audiobook.

PANEL 5: WORDS TO THINK ON

A Kindergartener's Way of Peace:
How to Better Achieve World Peace Through a Child's Morals

Rachel West

St. Joseph High School

Why the Phrase "Stop Acting Like a Child" Is Partly Invalid.

Adults and countries should be more like children and here is why. World peace would be best achieved by following the simple rules we tell young children such as: be kind, forgive and do not assume things about others. Children have no filter and thus they communicate even better than some adults being blunt and saying whatever comes to mind, while also taking in the response. Children's brains develop quickly, they are able to bounce back faster than older people, this makes it so when a child does something bad and gets told off by an adult, they change their behaviour (Rosenberg, 2020). That is why adults should make an effort to re-develop ideas even if it does not come as naturally as it does for children. Young children's perspectives on people are very simple, people who give them what they want are good and people who deprive them of that is bad, except despite the hyperbole commonly used by parents to describe their children's fighting, they in fact do not create wars when they don't approve of what happens. Although the reasoning behind war is not so black and white, we still need to understand that talking out problems is not something that only kids should do. Children show their emotions and most of the time will communicate what they want in some way, where countries end up in war to show what they want rather than making compromises or understanding what the other country is saying.

Babies and most young children do not care about the colour of each other's skin or what the other's ethnicity is. The innocence of a child makes it so they only care about who is it for "Tag" or who gets the last cookie (that being an exaggeration because kids care about more than just that). Inclusiveness within the world market would make it easier for world trade so people all around the world can get what they need. We expect children to follow our rules about kindness and sharing yet we ourselves do not do this. Between politics and fighting about "who deserves what" is irrelevant when we should be apologizing to the generations we have harmed and make amends so the countries of the world can have a clean slate to do better. Kids just want to have fun and not have to worry about life problems and by not having peace we are depriving them of their right to a good childhood.

We tell kids to be kind and to help each other pick up each other's messes, so why do we not do the same. Working in harmony with the rest of the world is the key to getting the world back on track, especially during these hard times involving the economy. Children start to show genuine empathy after they are around 2 years old, understanding how other people feel even when they don't feel the same way themselves (Abedon, 2005). If we had the empathy of children then maybe we could understand each other enough to live in peace. The innocence of kids is what keeps them impartial with how they treat others with basic courtesies and acts of kindness, which is also how everyone should treat one another. Let's start to show the compassion and kindness to each other that we want kids to show.

Why adults should accept others the same way children do before they fully develop. It does not matter if you are Catholic, an Atheist, part of the LGBTQ+ community, a women, a man, non-binary, Caucasian or not, we share this world and we should act like children in a sense that we should not discriminate. While kids have not fully developed into the people they are going to be, they can at least live-in moderate peace with each other and avoid making over the top, rude remarks. Kids are simple and while they may not get along all the time, they do not make extreme groups to combat important topics. Where adults almost always break into groups even for topics that both groups are right, or one is being petty. For example, pro-life verses pro-choice, pro-life is often genrally veiwed as a group who strongly believe in making abortions illegal, while pro-choice push for people with unplanned pregnancies, esspeaically among teenagers, when in relaity they can just co-exsist as options without pressuring people or trying to take away their options. We need to do better for the future generations, and it can start with the simplicity of being kind.

References

Booth, J. (2020, November 5). Are Children More Adaptable Than Adults? Experts Weigh In. Romper. https://www.romper.com/p/are-children-more-adaptable-than-adults-experts-weigh-in-40474839

Abedon, E. P. (2005, October 3). Toddler Empathy. Parents. https://www.parents.com/toddlers-preschoolers/development/behavioral/toddler-empathy/ d

The Backwardness of Online Education in the Rural Students of Bangladesh

Shimul Gupta[1]

[1]B.A (University of Liberal Arts Bangladesh) & M.A (University of Dhaka)

Abstract

In the field of education in Bangladesh, the digital learning process began with a new age. The Education Scheme in Bangladesh saw dramatic changes in 2020 with the onset of the coronavirus pandemic with the extensive use of online learning. Since the pandemic of Covid-19 in March 2020, all schools, colleges, and institutions in the country are closed, boosting the momentum for online education systems. Academic programming for Grade VI through Grade X pupils started to broadcast on national TV between 9:00 and 21:00 on March 28, 2020. In order to speed up online learning, the government and several State agencies supplied internet, cost-effective information, and smartphone credits. However, internet connectivity remained restricted in rural and semi-urban regions. Data from the Telecommunications Regulatory Commission (BTRC) from Bangladesh indicate that there are 93.7 million mobile internet users, 5.7 million internet broadband subscribers, and a tiny number of WiMAX customers in the nation. But Internet connectivity was not the only barrier that prevented students from benefiting from online education. Some of the main concerns which had still to be addressed were a shortage of qualified teachers, excellent material, a good environment, and financial assistance. The findings of this paper are the problems for leg-behind rural students in the usage of technical devices for online education.

Keywords: Online education, Bangladesh, COVID-19, Rural education, Academic program, etc.

Introduction

With nearly two hundred million inhabitants, the People's Republic of Bangladesh is one of the world's most densely populous countries within a region of 147,610 sq. kilometers. The massive population of the country is also the country's main resource. However, it's a good task for the govt. to show the people into a productive force and to confirm a dynamic setting for social, economic, and political growth. Though the official skill rate is claimed to be sixty-six percent, the verity rate is just forty-two percent consistent with personal surveys. Since Bangladesh became independent in 1971, education has been thought to be a keyspace for each administration. For a large number of people, Distance Education is the most significant difference for coaching in Bangladesh. (Sadeq, 2003). Most significantly, Bangladesh has severely limited possibilities for better education, and possibly even students, who are capable of funding their studies, discover that entrance to colleges because of low ability is highly difficult. The primary e-learning application or online training application in Bangladesh changed into e-learning or online teaching in 1956, and in 1992 Bangladesh created its first and handiest group for distance mastering referred to as the University of Bangladesh Open (BOU). The teaching era at the university is particularly conventional one-way, and students are listeners with examination papers. The decline fee in BOU is nevertheless pretty good-sized.

The Current Status of the Systems of Education in the Bangladesh

Four phases in Bangladesh are formal education: pre-primary education or early education; primary education (grade I to XII), secondary education, including two years of Bachelor's and three or four years of Bachelor's degree, (grades VI to XII, of which three first grades were considered junior or lower secondary, IX-X secondary and XI-XII secondary) The following images and tables illustrate the basic framework of different educational systems. Muslim students have a parallel system of formal religious instruction, which is known as Madrasah. For madrasah education, the equivalent of primary, secondary, higher secondary, bachelor, and master's is Ebtedayee, Dakhil, Alim, Fazil, and Kamil. In recent years' education in Madrasah has undergone a certain level of modernization and made the system closer to the mainstream school system in Bangladesh. Currently, about 3 million kids are engaged in education at Madrasah (ISLAM, 2007).

Bangladesh is mainly a homogenous country; however certain minority groups live in a variety of geographical locations. These groups comprise several ethnic groups from the divisions of Chittagong Hill Tract, Sylhet and Rajshahi, and Mymensingh. People in these places are distinct from Bangladeshi's dominant group languages, cultures, and geographical obstacles and the values of these people should become part of any inclusive education plan. Both the Government of Bangladesh and NGOs

Figure SEQ Figure* ARABIC 1
Education System in Bangladesh

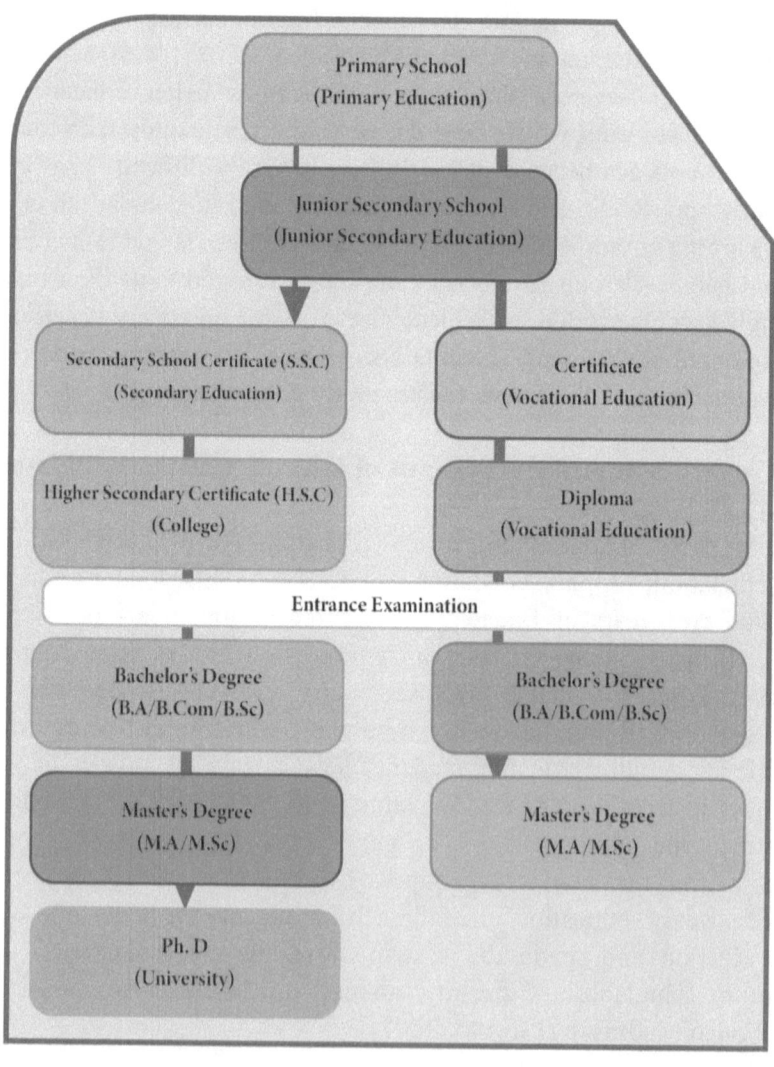

Note: **Basic Education Process in Bangladesh**

actively contribute to excellent elementary education and are involved in a number of specialized areas including inclusive education (Hase Ara Begum, 2018). However, history has shown that not all marginalized populations have benefited from inclusive education. As a country, Bangladesh is obligated to respect, protect and comply with the freedom to educate, without discrimination, in accordance with several international treaties on human rights (UDHR, ILO) Convention No 107, International Convention on Economic, Social and Cultural Rights (ICESCR) (Nations, 1948), United Nations Convention on Human Rights and the Rights of Human Rights (UN Convention) (ICCPR) (UN, 1957). Equitable rights for all people regarding gender, age, class, caste, ethnicity, and religion have also been provided by the Bangladesh Constitution.

For the current study, how government and NGO inclusive education efforts are carried out in rural and distant communities is particularly important. Geographic integration implies children in all parts of the country should be given parity, regardless of where the community is, but not always. Geographic inclusion in Bangladesh is covered by indigenous children, tea gardening kids, coastal children, hoar (wetland) kids, border children, among children, and the places impacted by flooding. In Chittagong hill, Sylhet, Mymensingh, and Patuakhali Indigenous groups are largely concentrated. The youngsters in the tea garden are in Sylhet, where tea is a significant business. South of Bangladesh, which are primarily disaster-prone areas, and Cox's Bazar are coastlines. The frontier territories cover Bangladesh's border districts such as Sathkhira, Khulna, Jessore, Thakurgaon, Sylhet, and Teknaf (McDonald, 2013). In the northern region of Bangladesh, there are Rangpur, Nilfamari, and Kurigram Mongolians, which are susceptible to hunger and famine. The geographical composition of Bangladesh, as can be seen from the above, is diverse. It follows that, in such a diversified terrain, isolated communities also have a range of cultural composites and would therefore provide inclusive education in specific fields. Research on how inclusive education in rural locations of developed nations is offered is rather recent. Much of Bangladesh's literature is in the form of government/NGO documents/policies. The article discusses how inclusive education can and is a continuing issue from the perspective of educators who work in the field.

Chittagong Hill Tract (CHT) Remote Indigenous children

The South-Eastern section of Bangladesh is the Chittagong Hill Tracts (CHT) area. With its geographical and distant variety, it differs from other areas of the country. Inclusive education has an influence on the traditional and cultural variety of the eleven various indigenous tribes, natural catastrophic emergencies, conflicts, and extended relocation. In terms of development indices, including access to excellent education, the area remains one of the most disadvantaged regions in the country. The CHT school enrollment percentage is relatively low, especially in distant rural regions. Under the UNDP-Chittagong Hill Tracts Development Facility Socioeconomic Baseline Survey CHT (CHTDF) (Barkat, 2009). The CHT school enrollment rate was 73%, which was below the national average of 99.4%. At 65 percent, the dropout rate was significant. The survey also found that just 7.8% of the households completed basic education and only 2.4% finished high school. About 54% of household managers did not have education; 9,4% finished elementary school, 4% finished high school and only 2% completed further education. For Bengali people, the number of school graduates (11%) in the CHT regions is more important than among indígena (8%). About 77 percent of women heads of the house had no education and 11 percent did not finish their elementary education but attended school. 44% of household members had no education in another survey done by the Manusher Jonno Foundation in 2011. 17% of households were under-literate. Primary school enrolment was 95%, but primary school drop-out rate was 59%. Junior school drop-out rates were 24%. Language difficulties, together with other problems such as remoteness, poverty, insufficient schools, the lack of indigenous local instructors and inadequate infrastructural facility, and a centralized national curriculum are described as one of the primary impediments to this poor performance of CHT education.

One of the major barriers to poor CHT education performance is the language challenges with other problems including distance, poverty, insufficient schools, absence of local indigenous instructors, inadequate infrastructural facilities, and a centralized national curriculum ((BBS), 2011). Under the National Education Policy for 2010, the MTB-MLE government has begun in Bangladesh with five main indigenous languages – Chakma, Marma, and Tripura in the CHT region, and Garo and Sadri in the plain countries – and is currently working in pre-primary schools in five major indigenous languages. Since the beginning of 2004 the Manusher Jonno Foundation (MJF), through local NGOs, has been conducting educational initiatives to secure the right to basic education and to enhance access to excellent education for all students. While the state plans to incorporate other groups in the MTB MLE Programme, the indigenous Santal community was not able to be included in the multilingual training program in the northern regions of the plain in 2017, since that community had lost its original script (has become extinct). The loss of local language in many rural regions is an increasing concern as we continue to pursue global educational aims. That we see above, there will be a little alternative for indigenous tribes to join the mainstream educational process if their future growth involves children in these fields.

Child Education in Flood Affected Areas

There is the lean season (September to November), every year until aman paddy (primary rice crop) is planted, where the poorest population, which mostly depends on the farm, remains unemployed. This is locally called a monga, which indicates that employment and food are very scarce. Children are compelled to leave school in these places as they are unable to eat. During monga, children can relocate to another place with parents or work for children (domestic in many cases) (Ahmed, 2005). Many kids of school age never go into school or go to work in these regions, which are subject to the season's food shortages (e.g. the yearly Monga Season). Research from a region where schools are caught indicated that the principal cause for not enrolling or leaving the school is 'scarcity of money or 'family poverty.' More than 40% of students who had dropped out were the major reason they dropped out of school.

Online Education System Before COVID-19 Situation

Online Education System at Secondary Education

Bangladesh government started in 2009 with the help from BRAC under TQI-SEP a pilot study on e-learning or online maths from Gazipur and Calla secondary schools. On 23 February 2010 TQI-SEP mobile ICT laboratory officially opened to enable e-learning of poor rural Bangladeshi high school pupils. Initially, many Mobile ICT laboratories have moved across the country to implement e-Learning systems for teachers and pupils from selective schools in 14 micro-buses and three four-wheel-drive trucks (for mountains, swampy areas, and isolated places). Each laboratory includes computers, WLAN, digital cameras, media projectors, webcams, printing machines, stylistic drives, interactive boards, e-learning CDs, speakers, generators, etc. The goal is to guarantee that students receive basic ICT knowledge and ICT-based education and increase teaching ability.

Online Assessment in Mathematics and Science

Daily Star and the Creative team jointly established the website and make effective use of the internet for Bangladeshi youngsters for school children named champs21.com. Champs21 is a web-based evaluation of mathematics and science for children in schools 3-10. Throughout the full session/year following the program, students can conduct chapter and term assessments. The application's adaptability aims to attain both theoretical and technical knowledge as well as relative clarity across students in rural areas.

Promoting Basic Education

This eLearning program was introduced in 2011 by Jago Foundation, a volunteer organization aimed at conducting 'online school' throughout the country. 10 online schools are now operated in 10 districts. A rural classroom with two local instructors is connected through the Internet and video conferencing to a teacher in Dhaka. This technique is as follows. Video conferencing software is used for interactive and professional purposes. Online teachers develop and manage audio/video clips, graphs, distribute test/homework tasks, monitor and evaluate the performance of the students, and collect data for analyzing progress. The instructors in the classroom use A / V equipment, keep attendance records, support students individually and in small groups in reinforcing online teacher learning ideas (Manna, 2015).

Online learning for English

Some non-governmental activities complement government activity. One of them was the D.Net (Development Research Network) CTEE project with funding assistance from VAB, New Jersey (US), which aims to examine the impact of e-learning of English on high school pupils. 10 schools have been taken to evaluate the efficiency of the created English learning CD. Two sets of pupils (Control Group and Experimental Group) were chosen from each of 10 schools (10 per group). The findings from the immediate final examinations showed similar outcomes. Therefore, a total of 200 experimental students have been chosen. The 'Supervisory Group' was Group 1 while the 'Experimental Group' was Group 2. An English test has been set for both groups. The response scripts were reviewed and rated with the aid of teachers such as 'Strong,' 'Medium' and 'Weak.' The control group included 100 kids (out of 10 schools), 61 in the weak group, 25 of them in the medium group and the remaining 14 of them in the strong group. The outcome, on the other hand, was weak-65, medium-22, and strong for the experimental group 13. A 36-hour session with 100 students utilizing the CD was done after the pre-test. In ordinary school classes, Multimedia Content was presented using a large-screen monitor. After six months of completing all classes at 10 schools, another exam for control groups as well as for experimental groups was taken. The control group got a normal English class alone. It is important to highlight. The after-test results revealed that multimedia learning had a substantial influence.

Impact of COVID-19 Solution in Education System

Primary school pupils are between 6 and 11 years of age. There are over 18 million students (USAID, 2020). During this period, most students suffer emotional pain from social differences. Poverty is growing among the lower class. The parents of the students become unemployed; their monthly payments mostly decrease. Because many private businesses have paid less than 50% during the COVID-19 epidemic (news, 2020). As a result, children are decreasing their capacity to buy nutritious food, and different mal-nutritional disorders are impacting their children.

They cannot tamper with their pals to keep their social distance, and cannot even play, which distresses memorabilia. Middle kids tend to be less attached to their parents. 91% of children and young persons expressed emotional pain and trouble according to the voices of children during the COVID-19 era. Mental disease is still a secret for youngsters, as they cannot communicate mental sickness with others. Two recent reports of low-income families in Barisal, Bangladesh have drastically decreased their monthly income (Alo, 2020). With less money, it's not feasible to educate their children and satisfy the demands of their children when food worth up to one month is not easy. In this period, parents have little access to safe and healthful meals and are economically and socially stressed. Parents are therefore more inclined, in the current scenario, to be with their children. According to the statements of the government of Bangladesh, there are more than 34 street children there (Independent, 2016). Most have had different facilities denied. However, several NGOs offer children free meals and other educational facilities. The COVID-19 epidemic has postponed its educational activities. They can therefore become involved in different criminal activities at a very young age. Sadly, COVID-19 pushes young children in Bangladesh to pray (Standard, 2020). The worst effect of Coronavirus on children is that they labor for child labor. On 7 April, however, government education was aired on SANGSAD TV (Express, 2020). However, numerous sources indicate that many families do not already have TV sets. Since schools are closed and pupils have delayed examinations for the first term, many elementary students believe that they don't grow up or have lost much because of the lack of schools, classes, and classes.

Based on the current secondary school department, there are about 23,000 high schools in Bangladesh (Board, 2020). In the study, we spoke with several high school kids aged between twelve and seventeen in our neighborhood. In their residence, 5 percent do not have a TV set. A severe familial regression affects 50 percent of the youngsters. 25% worry about higher education entrance. 25% worry. Although the administration has started television broadcasting classes, many pupils cannot, because many of these households do not have a TV. Recently, kids with an SSC are facing worry due to the postponement of college entrance certificates. As the students cannot begin their university sessions, they are quite concerned about shortening their academic year. Many students are obliged to work in a variety of dangerous tasks, for example constructing, clothing, driving the bus, and car rickshaws. Parents force girls at an early age to marry (Tribune, Is Covid-19 pandemic leading to a rise in child marriage? 2020). In addition, Bangladesh has a high early marriage rate (HRW, 2015). The female pupils are struggling to maintain their cleanliness throughout COVID-19. Many of them are therefore unable to use toilets because of lack of money. Many female students are forced by the poor economic conditions to work as sex workers to preserve family requirements.

Bangladesh has 9081 middle schools, and about a total of 13 lakhs (Tribune, 2019) are pupils. While the Higher Secondary (HSC) certificate test should be conducted, the COVID-19 epidemic postpones it. Students will be admitted to several institutions after completion of their secondary level and some will study more overseas. However, COVID-19 is hindering this. The pupils' future is now worrisome. Students that are capable of providing clever gadgets and internet access are glued to Facebook, Youtube, and online gaming. From the survey, numerous university students' parents, surprisingly, 60 percent of them are considered to be incorporated into mobile phones. You could not concentrate properly on your studies. Many don't read a single word at all. It's quite worrisome to most kids who waste their time with TV and social media or sit idly at-home quarantine. Many pupils are not prepared to take the HSC test since they are out of study already. Although the privately owned high schools are involved in online education, there are no online programs at government institutions that have yet begun. The study indicated that a large number of examined groups found both beneficial and useless online classes. It generates uncertainty in student decision-making. This imbalance may also occur in the governing body of some education institutions, the current study hypothesized. Motijheel Ideal School and College, Cambrian School and College, Viqarunnisa Noon School, Willes Little Flower School, College as well as several others are attending online classes (Tribune, 2020). There are 155 public and private universities in Bangladesh and 117 universities of healthcare and dentistry in Bangladesh (UGC, 2020), according to Bangladesh and Bangladesh Medical and Dental Council (BMDC) University Grant Commission (UGC). These institutes study more than 10 lakhs of pupils (BMDC, 2019). All medical and dental universities and universities remain closed since April 2020 during the pandemic scenario. Most students have visited their hometowns and communities. Some institutions take online lessons, although most of them left their electronic equipment at university or school dormitories, such as laptop desktops. Recently, the story was published by the English newspapers, which were polled from May 9 to May 11 by 2038 students from 42 public and private institutions. Only 23% of the total pupils are online. There are numerous reasons why you don't attend online classes. There's no broadband internet access available in village areas. If a student takes at least three courses each day, he or she must take 1 GB for an average length of up to 300 MB of data (IslamDMS, 2020). However, it is difficult to purchase for many students because of the increased price of Internet data. In genius and students in medicine face enormous difficulties. They cannot perform their workshops and practical lessons, which therefore have a detrimental effect on their careers.

Internet Connection Issues Hamper the Education System

The villages have a sluggish internet speed. The whole class participant is 58.8% from private universities, 41.2% from state universities. This makes it impossible for many of the students at public universities to acquire broadband Internet. Of the online scientific candidates as a whole, 55 percent, 12.1 percent, 11.2 percent, and 4.7 percent are studying Social Science from business studies and the rest of the field. The slowest data connection speed amongst 42 nations is said to be Bangladesh's daily star (DailyStar, 2020). We just got 7.8Mbps, when Canada has 63Mbps. It was also 9.2Mbps in early February, but it was 7.2Mbps in March. Most university students make money through part-time teaching or employment. With the shutdown of the stores and restaurants, many students lost their employment. Many parents have been unemployed because of the Coronavirus, or have become out of business. Some private institutions are also pushing students to pay semester fees. Students were concerned about how to deal with COVID-19's financial loss.

Impact of COVID-19 on Rural Education

During the school shutdown a widespread disturbance and significant loss in students' time due to COVID-19. It is normal to anticipate youngsters to spend more time studying at home in unplanned, long-term closure with private instructors and parents or utilizing technology for example, on television. However, the study time at home really fell during the shutdown of the school. This indicates that the loss of learning was higher than due to the close of school. In comparison to children from parents without education, the children of high school or higher (secondary education or above) were studying at home for longer. But the time spent on home learning even for the former group was far from filling up the gap in school loss (Asadullah, Bhattacharjee, Tasnim, & Mumtahena, 2020).

In order to compensate for school closures, the government provided lessons on national television. More than a third of the youngsters did not have access to a TV that was automatically dropped. Even when access was made available, relatively few kids watched television courses, such as just 25 percent of rural children who had access to television. Nearly every student polled did not use the Internet in their studying. Educating at home is not really an alternative for schools. The months of closing schools will probably make a profound, long-term impact on educational children's academic performance in Bangladesh. The government, therefore, has to develop remediations to compensate for the loss of learning during the epidemic.

Approximately 7 million (80% in rural regions) children and young people aged from 6 to 16 in 2016 were out of school in the state of Bangladesh. 87% of the population is women at an enormous disadvantage. The dropout rate at the primary and secondary levels, especially among children and girls from economically disadvantaged households, is anticipated to grow during current shutdowns. Although the frequency of dropouts in both elementary (about 18%) and secondary (about 35%) remained significant in recent years (Education, 2019). The failure of the second-chance pilot to return 1 million school children to schools would further deteriorate the situation. The predicted worldwide financial crisis caused by the pandemic will further raise the dropout rate, particularly for females, and for poor/ deprived households, as a result of the severe effect on household income.

An increasing number of school drops will probably lead to higher fertility and early marriage and child labor rates for young people.

Technical Problems Facing Bangladesh's Education System

Bangladesh now has 21.6 million primary and primary school pupils, another 13 million secondary school students and a further four million universities and institutions. 76% of secondary schools in Bangladesh are in rural regions, according to the Ministry of Education. It claims that almost 60 percent of pupils in elementary schools attend government-run institutions, especially in rural regions (A, 2020). Learning this manner offers a way of reducing communication between students or between students and professors. However, because of the absence of financial and digital resources or instruments, many students have no access to online education.

Obstacles of Online Learning Materials

The Internet is used as an instructing tool for educational institutions in Bangladesh. All levels, however, do not get identical rewards for pupils. Although the Internet is sufficiently fast for urban pupils, rural students remain trailing behind. Again, there is a divide between affluent and poor pupils owing to high internet costs and the unavailability of cell phones. Online lessons have become a benefit for students, based on Bangladesh's socioeconomic status. Everyone has different mental issues at the time of corona, including instability and frustration (Nuruddin Ahmed Masud, 2021). Online lessons have become a poison among them. In rural regions, students are not in a position to engage in online courses because of the poor quality of the internet in distant parts of the nation. They do not have sufficient energy, new cell phones, and even correct rules. In this period of a sluggish global pandemic, continuing education is a major issue. While you must purchase more data, Internet speed, and Wi-Fi, you must do it online, despite the problem of load shedding. The study rate is, nevertheless, diminishing progressively. Although the mobile and laptop type speed has risen, pupils will forget to write on the laptop.

Lack Experience of Teacher in Online Learning

The pandemic coronavirus is the cornerstone of online classrooms currently. But rural instructors are not yet comfortable with online teaching. Most students think that online education is distinct from physical education in the classroom, as they have never produced an online learning curriculum or have been exposed to an online learning curriculum. The primary difficulty is that the skill gap is faced by teachers in the field of online instruction. Most teachers have not received training on online learning approaches and technologies. Government must give teachers sophisticated technological training to continually upgrade their abilities (Nuruddin Ahmed Masud, 2021).

Adverse Environment During Online Learning in Home

There is a major distinction between classroom learning and home learning. Because the school setting is only appropriate for education when it is easy to learn in the environment, but on the other hand activity at home always takes place, movement of different people and various jobs, including homework, which is not entirely ideal for education (Nuruddin Ahmed Masud, 2021).

Conclusion

Bangladesh's large population is the country's main resource. However, the government continues to face major challenges by converting potential individuals into a productive force and guaranteeing a vibrant social, economic and political environment. It is officially reported that the literacy rate is 61%. Since independence, all administrations have consequently acknowledged education as a priority area. For various socio-economic reasons, distance education is a significant alternative to educate the masses in Bangladesh. More crucially, in Bangladesh, higher education opportunities are highly restricted, and it is thus very difficult to obtain admission to institutions even for those who can afford to fund their studies because of limited intake skills. In Bangladesh, there are unusually large drops in basic schooling through university level, mostly for economic and societal reasons.

References

(BBS), B. B. (2011). Population and Housing Consensus. Dhaka: Government of the People's Republic of Bangladesh.

A, P. (2020). COVID-19 impact on students. Crossref, 1-6.
Ahmed, M. a. (2005). Education Watch Report. Dhaka: Campaign for Popular Education (CAMPE).

Alo, P. (2020). Children at home in depression. Dhaka: Prothom Alo.

Asadullah, N., Bhattacharjee, A., Tasnim, M., & Mumtahena, F. (2020). Coronavirus Outbreak, Schooling and Learning: Study on Secondary School Students in Bangladesh. Dhaka: Brac University.

Barkat, A. (2009). Socio-Economic Baseline Survey of CHT. Dhaka: UNDP.

BMDC. (2019). New list of recognized medical & dental colleges and dental units (Govt. & Non-Govt.). Dhaka: BMDC.

Board, E. (2020). Dhaka: Ministry of education.

DailyStar, T. (2020). Mobile internets lowest in Bangladesh among 42 countries. Dhaka: DailyStar News.
Education, B. o.-F. (2019). Fourth Primary Education Program. Dhaka: MoPME.

Express, T. F. (2020). Primary students to get televised lessons through Sangsad TV. Dhaka.

Hasne Ara Beguma, R. P. (2018). The challenges of geographical inclusive education in rural Bangladesh. INTERNATIONAL JOURNAL OF INCLUSIVE EDUCATION.

HRW. (2015). Bangladesh: Girls damaged by child marriage. Dhaka: HRW.
Independent, T. (2016). Street children in Bangladesh:A life of uncertainty. Dhaka: The Independent.

ISLAM, M. R. (2007). The role of education for rural population transformation in Bangladesh. Asia-Pacific Journal of Cooperative

Education, 8(2), 1-21.

IslamDMS, T. A. (2020). Online classes for university students in Bangladesh during the Covid-19 pandemic: is it feasible? Dhaka: Tbs news.

Manna, D. M. (2015). E-LEARNING IN BANGLADESH: COUNTRY REPORT. Seoul: Seoul Cyber University.

McDonald, L. a.-D. (2013). Moving Forwards, Sideways or Backwards? Inclusive. International Journal of Disability, Development and Education, 70-84.

Nations, U. -U. (1948). Universal Declaration of Human Rights. Paris: UN. news, T. (2020). Here comes the pay chop. Tbs news.

Nuruddin Ahmed Masud, T. N. (2021). Impact of online Learning during Covid-19: A Study of Rural Area in Bangladesh. International Journal of Creative Research Thoughts (IJCRT).

Sadeq, A. M. (2003). Cooperation and collaboration for ODE: The case of Bangladesh. 17th AAOU Annual Conference. Thailand.

Standard, T. B. (2020). Covid-19makingthingsworseforchildren. Dhaka: Tbs news.

Tribune, D. (2019). HSC, equivalent exams results: Higher pass rates, more GPA-5 achievers this year. Dhaka: Dhaka Tribune.

Tribune, D. (2020). Covid-19: Educational institutions engaging in online,virtual classes. Dhaka: Dhaka Tribune.

Tribune, D. (2020). Is Covid-19 pandemic leading to a rise in child marriage? Dhaka: Dhaka Tribune.

UGC. (2020). Public Universities. Dhaka: UGC.

UN, U. N. (1957). International Labour Organization (ILO) Convention No 107. NewYork: UN.

USAID. (2020). EDUCATION. USAID.

BIOGRAPHIES

The following entries are from conference presenters who were unable to publish their articles.

Slowing the Sands of Time.

Thomas Banks is a Political Science student at the University of Alberta and a 2021 intern with AIC.

Objections to Immortality.

Jonathan Wiebe. This past year he finished a bachelor's degree in Psychology at the University of Alberta, and will be going into his first year of Yorkville University's Counselling Psychology Masters program this fall.

The Impact of COVID-19 on Ocean Acidification and What This Means for Humans.

Amal Rizvi is a fourth year undergraduate student, currently enrolled in the honours biology program under Western University's faculty of science. She's passionate about the intersection between environmental systems with human and animal health.

The Voynich Manuscript: (Un)covering our Mysterious Past.

Samira Sunderji is a fourth year Honours Kinesiology student at McMaster University, and is highly engaged with her academic community. This is her second AIC Conference presentation, and she is looking forward to learning more from colleagues, networking with others, and furthering her passions for writing and education. Her presentation divulges into the research behind one of her recently published books and engages with how mysterious our past truly is.

Is This Funny? A Generic Framework for Critically Reading North American Stand-Up Comedy as Text Using Literary Theories.

Elisia Snyder is a sessional instructor at MacEwan University and works with the University of Alberta Press. In 2017 she received a SSHRC CGS-M Grant to fund her masters. Elisia also works as a touring stand-up comedian and has performed at the Okanagan Comedy Festival.

Beatles, Music, Storytelling, Deeper Meaning, I am the Walrus, Glass Onion.

Kyra Droog is a graduate of MacEwan University's Bachelor of Communication Studies program. She is a nine-time published author, with three short fiction credits and six non-fiction credits. Kyra has worked in a variety of fields, including recreation, communication, editing, and publishing, and is passionate about the power that comes from reading and writing.

Rocky Raccoon and Bungalow Bill: American Music's Influence on the Beatles Songwriting.

Ryan McMillen is a MacEwan University Bachelor of Communications graduate with a deep interest in how media of all forms influences us. A former oilfield worker, he also has a particular interest in workplace structure and culture. Now a working communications professional, Ryan continues to explore his fascination with media and workplaces through his own research and writing.

Effectiveness of Global Rural Teleoncology and Future Routes for Teleoncology Research in the Context of Canadian Rural Areas: A Narrative Review of the Literature.

Author(s): Parmin Rahimpoor-Marnani (School of Kinesiology and Health Science, York University), David Vaz (Faculty of Health Sciences, McMaster University), Jasrita Singh (Department of Biochemistry and Biomedical Sciences, McMaster University), Daivat Bhavsar (Faculty of Health Sciences, McMaster University), Vivek Kannan (Faculty of Engineering, McMaster University), Austin Mardon (Department of Psychiatry, University of Alberta). Parmin Rahimpoor-Marnani is entering her 4th year of undergraduate studies in Kinesiology and Health Science at York University. She is particularly interested in the systematic and socioeconomic factors that impact the accessibility of healthcare services. Parmin is a volunteer with AIC's Sharpen the Quill program and a research assistant at two neuroscience laboratories at York University. In her free time, she enjoys reading and running. This article was written as part of AIC's "Sharpen the Quill" program and is currently under review with a chance of publication at the University of Toronto Medical Journal (UTMJ).

Prevalence and Adverse Effects of Mental Illness: Depression and Anxiety within Canadian South Asian Immigrants and Children of Immigrants.

Dhwani Bhadresa is entering her fourth year of a Bachelors of Kinesiology, with a minor in Sociology at McMaster University. Dhwani is passionate about serving the community; she's held leadership positions in McMaster Indian Association and is the co-founder of the Paediatric Mental Health Initiative. Her interest and research is focused on analyzing mental health within the South Asian community in order to facilitate tangible change.

From Childhood Abuse to Anorexia Nervosa: The Roles of Aberrant Stress Signalling, Dissociation, and Emotional Avoidance.

Lydia Sochan is a third year biochemistry student at the University of Western Ontario. Lydia has done clinical work with children at the University of Washington Autism Centre, and research in brain aging through the OwenLab at UWO. She loves to combine her knowledge of the biological sciences with psychology and she has a particular fascination for the biological factors involved in the etiology and maintenance of conditions such as anorexia nervosa and autism.

Mistborn's Role in Generating Philosophical Debate.

The author of the essay "Mistborn's Role in Generating Philosophical Debate" is Elise West, a young writer just starting at Grant Macewan University. She is an avid fan of fantasy novels, which is what spawned her idea to write about one of her favourite book series, Mistborn. She expresses her view on how Mistborn is a prime example of how fantasy may encourage open-mindedness in its readers, more specifically, about how Mistborn encourages people to talk about religion, rather than shy away from it as if it is a taboo topic. Her paper was presented in the 2021 AIC conference, which was recorded and uploaded to the Antarctic Institute of Canada Youtube channel. The paper can be found in the published anthology titled Religious Themes in Speculative Fiction Texts.

www.ingramcontent.com/pod-product-compliance
Lightning Source LLC
Chambersburg PA
CBHW020234030726
47497CB00009B/3095